Occurrence Survey of Pharmaceutically Active Compounds

About the Awwa Research Foundation

The Awwa Research Foundation (AwwaRF) is a member-supported, international, nonprofit organization that sponsors research to enable water utilities, public health agencies, and other professionals to provide safe and affordable drinking water to consumers.

The Foundation's mission is to advance the science of water to improve the quality of life. To achieve this mission, the Foundation sponsors studies on all aspects of drinking water, including supply and resources, treatment, monitoring and analysis, distribution, management, and health effects. Funding for research is provided primarily by subscription payments from approximately 1,000 utilities, consulting firms, and manufacturers in North America and abroad. Additional funding comes from collaborative partnerships with other national and international organizations, allowing for resources to be leveraged, expertise to be shared, and broad-based knowledge to be developed and disseminated. Government funding serves as a third source of research dollars.

From its headquarters in Denver, Colorado, the Foundation's staff directs and supports the efforts of more than 800 volunteers who serve on the board of trustees and various committees. These volunteers represent many facets of the water industry, and contribute their expertise to select and monitor research studies that benefit the entire drinking water community.

The results of research are disseminated through a number of channels, including reports, the Web site, conferences, and periodicals.

For subscribers, the Foundation serves as a cooperative program in which water suppliers unite to pool their resources. By applying Foundation research findings, these water suppliers can save substantial costs and stay on the leading edge of drinking water science and technology. Since its inception, AwwaRF has supplied the water community with more than $300 million in applied research.

More information about the Foundation and how to become a subscriber is available on the Web at **www.awwarf.org**.

Occurrence Survey of Pharmaceutically Active Compounds

Prepared by:
David L. Sedlak and **Karen Pinkston**
Department of Civil and Environmental Engineering
University of California, Berkeley, CA 94720
and
Ching-Hua Huang
School of Civil and Environmental Engineering
Georgia Institute of Technology, Atlanta, GA 30332

Jointly sponsored by:
Awwa Research Foundation
6666 West Quincy Avenue, Denver, CO 80235-3098
and
WateReuse Foundation
635 Slaters Lane, 3rd Floor, Alexandria, VA 22314

Published by:

and

DISCLAIMER

This study was jointly funded by the Awwa Research Foundation (AwwaRF) and the WateReuse Foundation (WRF). AwwaRF and WRF assume no responsibility for the content of the research reported in this publication or for the opinions or statements of facts expressed in the report. The mention of trade names for commercial products does not represent or imply the approval or endorsement of AwwaRF or WRF. This report is presented solely for informational purposes.

Printed in the U.S.A.

ISBN 1-58321-370-8

Printed on recycled paper

CONTENTS

TABLES

FIGURES

FOREWORD

The Awwa Research Foundation is a nonprofit corporation that is dedicated to the implementation of a research effort to help utilities respond to regulatory requirements and traditional high-priority concerns of the industry. The research agenda is developed through a process of consultation with subscribers and drinking water professionals. Under the umbrella of a Strategic Research Plan, the Research Advisory Council prioritizes the suggested projects based upon current and future needs, applicability, and past work; the Collaborative Research, Research Application, and Tailored Collaborations programs; and various joint research efforts with organizations such as the U.S. Environmental Protection Agency, the U.S. Bureau of Reclamation, and the Association of California Water Agencies.

This publication is a result of one of these sponsored studies, and it is hoped that its findings will be applied in communities throughout the world. The following report serves not only as a means of communicating the results of the water industry's centralized research program but also as a tool to enlist further support of the nonmember utilities and individuals.

Projects are managed closely from their inception to the final report by the foundation's staff and a large cadre of volunteers who willingly contribute their time and expertise. The foundation serves a planning and management function and awards contracts to other institutions such as water utilities, universities, and engineering firms. The funding for this research effort comes primarily from the Subscription Program, through which water utilities subscribe to the research program and make an annual payment proportionate to the volume of water they deliver and consultants and manufacturers subscribe based upon their annual billings. The program offers a cost-effective and fair method for funding research in the public interest.

A broad spectrum of water supply issues is addressed by the foundation's research agenda: resources, treatment and operations, distribution and storage, water quality and analysis, toxicology, economics, and management. The ultimate purpose of the coordinated effort is to assist water suppliers to provide the highest possible quality of water economically and reliably. The true benefits are realized when the results are implemented at the utility level. The foundation's trustees are pleased to offer this publication as a contribution toward that end.

Walter J. Bishop
Chair, Board of Trustees
Awwa Research Foundation

James F. Manwaring, P.E.
Executive Director
Awwa Research Foundation

ACKNOWLEDGMENTS

The authors of this report are indebted to the following water utilities and individuals for their cooperation and participation in this project:

R.M. Clayton Water Reclamation Center, Atlanta, GA
Dublin/San Ramon Services District, Dublin, CA
West and Central Basin Water District, Los Angeles, CA
Mt. View Sanitation District, Martinez, CA
City of San Jose Water Pollution Control Plant, San Jose, CA
South Cobb Wastewater Treatment Plant, Cobb County, GA
F. Wayne Hill Water Resources Center, Gwinnett County, GA
Sweetwater Recharge Facility, Tuscon, AZ
Professor Robert Arnold and his students at the University of Arizona

In addition, the authors are very grateful to the Project Advisory Committee (PAC), Mark Burkhardt, David Lipsky, Lucia McGovern, Kimberly Ajy and Thomas Heberer, for their advice and assistance. The authors also express their gratitude to Djanette Khiari, the AwwaRF project manager, for her patience and support.

EXECUTIVE SUMMARY

Pharmaceutically active compounds (PhACs) are a family of compounds that includes prescription drugs, over-the-counter medications, drugs used in hospitals and veterinary drugs. Starting in the 1990s, a variety of PhACs were reported in wastewater effluent, surface waters and groundwater in Europe. In a few isolated cases, PhACs also were detected in drinking water. Although no known human health effects have been associated with exposure to drinking water containing trace concentrations of PhACs, members of the scientific community, regulators and consumers have expressed concerns about the presence of PhACs in water supplies. Water supplies on the United States are subject to many of the same potential sources of contamination as those studied in Europe and much less information is available on the occurrence and fate of PhACs in the US.

The concentrations of PhACs detected in the aquatic environment typically are less than 100 ng/L (i.e., <0.1 μg/L). As a result of the difficulties associated with measuring trace concentrations of PhACs, it is extremely difficult to detect PhACs in most water supplies, even when the compounds are present in a water supply. Because most PhACs originate in wastewater effluent, water supplies that partially consist of wastewater effluent are the most likely locations where PhACs will be detected. In particular, it is most likely that PhACs will be detected in locations where indirect potable water reuse is practiced.

RESEARCH OBJECTIVES

The American Water Works Association Research Foundation (AwwaRF) funded this study to assess the occurrence of PhACs in drinking water supplies in the United States (US). This report presents the results of the research. Specific objectives of the research included:

1. Evaluation of available information on the use, occurrence and analysis of PhACs that are potentially present in water sources in the US.
2. Selection of PhACs that are likely to be present at detectable concentrations in municipal wastewater effluent and agricultural runoff in the US.
3. Development of robust and simple analytical methods for quantifying PhACs in the aquatic environment.
4. Analysis of samples from sites that are likely to contain elevated concentrations of PhACs.
5. Preliminary assessment of the ability of advanced wastewater treatment plants, engineered treatment wetlands and soil aquifer treatment systems to remove PhACs.

APPROACH

In the first phase of the project, the compounds that were most suitable for monitoring were identified. The process used to identify candidate PhACs for monitoring included an evaluation of the use patterns of prescription drugs and their metabolism, a review of occurrence data from Europe and a review of analytical methods. After identifying compounds to be monitored, analytical methods were developed and tested. The methods then were used to analyze samples from conventional and advanced wastewater treatment plants, soil aquifer treatment systems, engineered treatment wetlands and surface waters. Preliminary studies also were conducted to assess the role of chlorine disinfection in the transformation of PhACs.

CONCLUSIONS

1. Although a large number of PhACs are used in the US, a relatively limited number of compounds can be detected frequently with existing analytical methods.
2. PhACs can be monitored by gas chromatography/tandem mass spectrometry (GC/MS) and high performance liquid chromatography/mass spectrometry (HPLC/MS) but the analytical methods require careful attention to detail and rigorous quality assurance/quality control (QA/QC) measures.
3. Acidic drugs, beta-blockers and antibiotics often are present in the effluent of conventional municipal wastewater treatment plants at concentrations between 10 and 10,000 ng/L.
4. PhACs appear to be removed effectively in advanced wastewater treatment plants equipped with reverse osmosis or granular activated carbon. However, trace concentrations of PhACs were detected in effluent from advanced treatment plants on two occasions.
5. Most PhACs appear to be removed during soil aquifer treatment. However, low concentrations of certain PhACs can be detected after soil aquifer treatment.
6. Little removal of PhACs occurs in the two engineered treatment wetlands studied, even at hydraulic residence times of up to one week.
7. Chlorine disinfection of wastewater effluent may result in transformation of certain PhACs, provided that ammonia is not present in the wastewater effluent. The transformation is usually incomplete under the conditions encountered during wastewater disinfection. These same reactions also can occur in sample containers and care must be exercised when collecting samples that contain chlorine residual.

RECOMMENDATIONS

1. Although there are no known human health effects associated with exposure to PhACs, utilities should be aware that PhACs are likely to be present in source waters that receive wastewater effluent from conventional municipal wastewater treatment plants.
2. If it is necessary to monitor PhACs in drinking water sources, the analytical methods described in this report can be used. However, it is important for the laboratory to analyze control samples and matrix recovery samples before beginning a monitoring program.

FUTURE RESEARCH

The results of this study and several other recent studies confirm the presence of PhACs in drinking water sources in the US. To assess the potential risks associated with these compounds, additional research is needed on the occurrence of PhACs and the efficacy of different types of treatment employed in advanced wastewater treatment plants. Research also is needed on the contributions of PhACs from sources unrelated to wastewater effluent, such as animal feeding lots and leaking sewers. This research would be enhanced by the development of additional sensitive and robust analytical techniques. In addition, research is needed on the removal of PhACs in receiving environments, such as soil aquifer treatment systems, bank filtration systems and engineered treatment wetlands.

CHAPTER 1
BACKGROUND AND LITERATURE REVIEW

BACKGROUND

Pharmaceutically active compounds (PhACs) are a group of compounds consisting of prescription drugs, over-the-counter medications, drugs used exclusively in hospitals and veterinary drugs. Many PhACs are excreted in their original form or as sulfate or glucuronide conjugates. Many of these conjugates can be converted back into the active parent compound by enzymes present in wastewater. As a result of the relatively large quantities of drugs used worldwide, municipal wastewater and animal wastes contain detectable concentrations of a number of PhACs. Starting in the 1990s, scientists from Europe began reporting the detection of a variety of different PhACs in wastewater effluent and in effluent-impacted surface waters. In several isolated cases, PhACs also were detected at low concentrations in drinking water.

When this project was initiated, during the spring of 2000, little information was available on the occurrence of PhACs in potential drinking water sources in the United States. However, the similarities between wastewater treatment systems in the US and Europe led scientists to hypothesize that PhACs also would be present in wastewater effluent and that drinking water supplies in the US that are subject to wastewater effluent discharges also might contain PhACs. However, drug use patterns in the US are different from those in Europe, and it was unclear which compounds would be the most important PhACs in the US. The main objective of this research project was conducted to identify the PhACs that could be present in water supplies in the US.

To identify PhACs of significance to water suppliers, the first step in the project involved a literature review and a review of available data on the occurrence of PhACs. The literature review was designed to identify a group of compounds to be measured as part of the occurrence survey. The review addressed the quantities of prescription drugs administered in the US, previously published studies from around the world that have documented the occurrence of PhAcs in water supplies and the analytical methods that can be used to measure PhACs. The selection criteria for compounds to be studied during the occurrence survey, which were based upon results of the literature review included their expected concentrations in water supplies, environmental fate, potential effects and the availability of suitable analytical methods. Since use patterns for PhACs change rapidly, and it is time consuming and expensive to measure all of the PhACs potentially present in the aquatic environment, the review was focused on compounds that can be used as representatives or indicators in future studies.

After identifying the PhACs that would be most suitable for study, analytical methods were developed and tested. In most cases, the analytical methods were modified from previously published methods. However, most of the previously published methods were complicated and could not be used exactly in the manner described in publications due to the lack of availability of certain standards or difficulties associated with replication of published protocols. Therefore, the analytical methods were modified and simplified to provide convenient and robust techniques that could be used by utility or commercial laboratories interested in monitoring PhACs.

After completing the method development and testing phase of the project, a limited occurrence survey consisting of 18 locations sampled between one and three times each was conducted for the targeted PhACs. The purpose of the occurrence survey was to determine concentrations of PhACs in drinking water sources that were most likely to contain the target compounds. As a result, the study focused on wastewater effluent and sites in which indirect potable water reuse was practiced. The occurrence survey did not attempt to cover water supplies in which wastewater effluent was not a major component. Therefore, the results of the occurrence survey are limited and likely biased towards situations related to intentional or unintentional water reuse.

LITERATURE REVIEW

Several excellent literature reviews recently have been published on the occurrence of PhACs in the aquatic environment (Daughton and Ternes 1999, Kummerer 2001, Heberer 2002, Snyder et al. 2003). These reviews summarize the occurrence of different PhACs in wastewater, the analytical methods used to measure the compounds, the efficacy of different treatment methods and the available data on the effects of PhACs on aquatic organisms. Rather than replicating these existing reviews, we reviewed the scientific literature to obtain additional information on the expected concentrations of PhACs in wastewater effluent and the applicability of candidate analytical methods. For practical reasons, the review is separated into sections on drugs used in human therapy and drugs used in animal husbandry, which mainly consist of antibiotics.

Drugs Used in Human Therapy

To estimate the concentrations of PhACs in municipal wastewater, pharmaceutical industry data were considered for the most popular prescription drugs in the United States. Information on other PhACs likely to be present in municipal wastewater (e.g., over-the-counter drugs, drugs used exclusively in hospitals) was obtained by reviewing occurrence data for PhACs in municipal wastewater effluent. The first step in selecting PhACs for further study involves estimating the concentrations of compounds present in water discharged by municipal wastewater treatment plants. To achieve this objective, quantitative estimates were made for the most popular US pharmaceuticals in municipal wastewater. Estimates were not made for less popular drugs, non-prescription drugs, and drugs used mainly in hospitals (e.g., X-ray contrast media, cancer chemotherapy drugs) because sales data were unavailable. Discussions with pharmaceutical industry consultants indicated that information on less popular prescription drugs could be purchased at considerable cost from proprietary databases. Since our intention was to identify the most important PhACs present in US waters, we chose to rely upon occurrence data to identify important PhACs not included in the survey.

Our approach for estimating the concentrations of prescription drugs in untreated wastewater involved dividing the mass of drug excreted by patients by the volume of wastewater discharged to municipal wastewater treatment plants. Calculations were performed for the top 200 prescription drugs listed in a 1998 survey conducted by IMS Health (1999). The top 200 prescription drugs included a total of 136 PhACs, because some of the drugs contain the same active ingredient. Because numerous assumptions were made to convert the number of prescriptions administered to the concentration of PhACs in municipal wastewater, considerable uncertainty is associated with these estimates. Despite the uncertainties, the estimates were useful in identifying PhACs that were candidates for further study.

Estimation of the concentrations of PhACs in municipal wastewater required the conversion of the number of prescriptions administered into the mass of compound discharged. Because several formulations are available for each prescription drug, the mass of active ingredient in a dose varies between prescriptions. For example, the beta-blocker timolol is prescribed at 40 mg/prescription in an oral formulation and 6 mg/prescription in an eye ointment. To estimate the mass of active PhAC associated with each dose, we consulted medical reference books (Katzung 1998, PDR 1999) and interviewed a practicing pharmacist who provided information on the most popular form of each prescription of interest (Field 1999). After estimating the mass of active ingredient in each dose of the most popular form of the drug, we estimated the number of doses per prescription. Estimates were made for the maximum and minimum masses per prescription assuming the most common drug formulations. For drugs that were given on a one-time basis (e.g., antibiotics), we assumed that each prescription included a sufficient number of doses to treat the ailment (typically 10 days). For drugs administered on a continuing basis (e.g., beta-blockers, birth control pills) we assumed that each prescription was renewed monthly. The basis for this assumption was the current practices of many health maintenance organizations (HMOs) to refill prescriptions once per month (Field 1999).

After estimating the mass of each drug prescribed, we estimated the concentration present in untreated wastewater. We assumed the population of the US is 250 million, that each person produces 320 L of wastewater per day and that the use of prescription drugs is equally distributed across the country. When excretion data were readily available, we estimated the fraction of the dose excreted in its original form. However, excretion data were not readily available for many drugs, or when the data were available, it was unclear if glucuronide or sulfate conjugates were considered to be transformation products. Since the conjugates appear to be converted back into their original, unconjugated forms prior to, or during, wastewater treatment, conjugated forms of drugs should be included with the PhACs. As a result of missing or ambiguous data, information on metabolism was only available for 30% of the PhACs considered. Therefore, comparisons between estimated concentrations of PhACs are made without consideration of metabolism. No attempts were made to quantify pharmaceutically-active metabolites. Additional details of these calculations are provided in Pinkston (2004).

Estimated concentrations of prescription drugs in untreated wastewater (Table 1.1) ranged from less than 1 ng/L to approximately 133,000 ng/L. The estimated concentrations were distributed over a wide range, with the majority of compounds estimated to be present at concentrations between 100 and 1,000 ng/L (Figure 1.1). In general, the compounds expected to be present at the highest concentrations consisted of pain relievers (e.g., acetominophen, ibuprofen) and antibiotics (e.g., cephalexin, amoxicillin). Because some of the pain relievers on the list also are available as over-the-counter products, their concentrations in wastewater could be considerably higher. Compounds estimated to be present at the lowest concentrations tended to be potent drugs such as hormones (e.g., medroxyprogresterone, equilin). Therefore, compounds estimated to be present at low concentrations are not necessarily unimportant. However, they would be difficult to detect without extremely sensitive analytical methods.

Table 1.1

Geometric mean for range of estimated concentrations of PhACs in municipal wastewater in the United States

		Predicted concentration excluding metabolism (ng/L)		Predicted concentration including metabolism (ng/L)		
Name	Classification	Geometric mean	Predicted range	Geometric mean	Predicted range	Excretion*
acetominophen	analgesic	61,000	29,000 to 130,000	53,000	25,000 to 111,000	C
ibuprofen	analgesic, anti-inflammitory	37,000	30,000 to 45,000			B
amoxicillin	antibiotic	27,000	19,000 to 38,000	16,000	11,000 to 23,000	D
metformin	antidiabetic	24,000	14,000 to 41,000	21,000	12,000 to 37,000	D
cephalexin	antibiotic	14,000	6,800 to 27,000	12,000	6,100 to 24,000	D
nabumetone	analgesic, anti-inflammitory	12,000	8,300 to 17,000			F
azithromycin	antibiotics	9,200	1,400 to 61,000			J
oxaprozin	analgesic, anti-inflammitory	6,600	6,600			B
sodium valproate	anticonvulsant	6,000	1,700 to 21,000	2,400	700 to 8,400	H
gabapentin	anticonvulsant	5,400	3,800 to 7,600	5,400	3,800 to 7,600	D
carisoprodol	skeletal muscle relaxant	5,100	5,100			J
penicillin	antibiotic	4,000	3,100 to 5,200			J
sulfamethoxazole	antibiotic	3,800	1,700 to 8,400	3,200	1,400 to 7,200	G
gemfibrozil	cholesterol lowering	3,400	3,400			D
metoprolol	β-blocker	3,100	1,500 to 6,600	160	74 to 330	B
ciprofloxacin	antibiotic	3,100	2,200 to 4,300	1,400	970 to 1,900	G
ranitidine	H_2-receptor antagonist	3,000	2,100 to 4,200			J
mupirocin	antibiotic	2,800	2,000 to 4,000			A

(continued)

Table 1.1 (Continued)

Name	Classification	Predicted concentration excluding metabolism (ng/L)		Predicted concentration including metabolism (ng/L)		Excretion*
		Geometric mean	Predicted range	Geometric mean	Predicted range	
clarithromycin	antibiotic	2,800	2,000 to 3,900	680	390 to 1,200	F
phenytoin	anticonvulsant	2,700	2,300 to 3,100			A
diltiazem	calcium channel blocker	2,600	1,500 to 4,500	79	45 to 140	A
naproxen	analgesic, anti-inflammitory	2,400	1,700 to 3,500			C
verapamil	calcium channel blocker	2,400	1,500 to 4,000	85	52 to 140	J
ipratropium	bronchiodiolater	2,400	2,400			E
trimethoprim	antibiotic	2,200	670 to 7,100	1,500	450 to 4,700	J
tramadol	analgesic	2,200	1,100 to 4,300	640	320 to 1,300	E
cimetidine	H_2-receptor antagonist	2,200	1,100 to 4,300	1,000	510 to 2,100	D
clavulanic acid	antibiotic	2,100	2,100	680	680	E
propoxyphene	opiod analgesic	2,100	970 to 4,500			J
bupropion	antidepressant	2,100	1,700 to 2,600			F
hydrochlorothiazide	diuretic	1,900	1,400 to 2,600	1,200	830 to 1,600	D
troglitazone	antidiabetic	1,800	1,000 to 3,100			B
cefprozil	antibiotic	1,700	1,200 to 2,400	1,000	710 to 1,400	J
pseudoephedrine	decongestant	1,600	1,600			J
erythromycin	antibiotic	1,500	1,500	75	75	I
atenolol	β-blocker	1,500	520 to 4,100	1,500	520 to 4,100	D
sertraline	antidepressant	1,400	490 to 3,900	180	64 to 510	F
triamterene	diuretic	1,400	1,400			J
nefazodone	antidepressant	1,300	750 to 2,200	13	7.5 to 22	F

(contined)

Table 1.1 (Continued)

Name	Classification	Predicted concentration excluding metabolism (ng/L)		Predicted concentration including metabolism (ng/L)		Excretion*
		Geometric mean	Predicted range	Geometric mean	Predicted range	
tetracycline	antibiotic	1,200	830 to 1,700			J
allopurinol	antigout	1,000	600 to 1,800			F
furosemide	diuretic	960	490 to 1,900			J
cefuroxime	antibiotic	900	450 to 1,800	900	450 to 1,800	D
nizatidine	H_2-receptor antagonist	860	610 to 1,200	520	370 to 730	D
fluoxetine	antidepressant	860	490 to 1,500			F
omeprazole	antiulcerative	850	480 to 1,500			F
amitriptyline	antidepressant	850	480 to 1,500			F
nifedipine	calcium channel blocker	760	530 to 1,100			J
codeine	opiod analgesic	730	230 to 2,300			C
trazodone	antidepressant	620	390 to 1,000			F
atorvastatin	cholesterol lowering	630	220 to 1,800	12	4.4 to 35	I
lisinopril	ACE Inhibitor	590	290 to 1,200	590	290 to 1,200	F
losartan	antihypertensive	560	320 to 970	23	13 to 39	F
loracarbef	antibiotic	480	340 to 680			J
fluconazole	antifungal	480	240 to 950			J
fexofenadine	antihistamine	470	470			J
paroxetine	antidepressant	440	180 to 1,100	9	3.6 to 22	F
valsartan	antihypertensive	440	220 to 880			J
levofloxacin	antibiotic	400	280 to 570	350	250 to 490	D

(continued)

Table 1.1 (Continued)

Name	Classification	Predicted concentration excluding metabolism (ng/L)		Predicted concentration including metabolism (ng/L)		
		Geometric mean	Predicted range	Geometric mean	Predicted range	Excretion*
cisapride	gastroprokinetic	380	270 to 530	38	27 to 53	D
nitrofurantoin	antibiotic	350	250 to 500	130	90 to 180	E
loratadine	antihistamine	340	340	34	34	J
famotidine	H_2-receptor antagonist	320	160 to 630	85	43 to 170	J
venlafaxine	antidepressant	300	210 to 420	15	10 to 21	F
isosorbide dinitrate	antianginal	280	280			F
quinapril	ACE Inhibitor	270	94 to 760			F
hydrocodone	opiod analgesic	270	98 to 720			J
propranolol	β-blocker	250	89 to 710			J
cyclobenzaprine	skeletal muscle relaxant	240	240			J
prednisone	glucocorticoid	240	69 to 830			F
enalapril	ACE Inhibitor	240	84 to 670	240	84 to 670	F
pravastatin	cholesterol lowering	240	120 to 470			I
simvastatin	cholesterol lowering	230	82 to 650			I
benazepril	ACE Inhibitor	220	110 to 460			F
fluvastatin	cholesterol lowering	180	130 to 250			I
Lovastatin	cholesterol lowering	180	88 to 350			I
nisoldipine	calcium channel blocker	150	110 to 210	15	11 to 21	B
glipizide	antidiabetic	130	60 to 290	13	6.0 to 29	A
fosinopril	ACE Inhibitor	110	54 to 220			F

(continued)

Table 1.1 (Continued)

Name	Classification	Predicted concentration excluding metabolism (ng/L)		Predicted concentration including metabolism (ng/L)		
		Geometric mean	Predicted range	Geometric mean	Predicted range	Excretion*
zafirlukast	antiasthmatic	100	100			J
promethazine	antihistamine	91	63 to 130			J
tamoxifen	antiestrogen	79	57 to 110			F
sildenfil		79	57 to 110			J
warfarin	anticoagulant	80	36 to 180			J
medroxyprogesterone	hormone	65	51 to 82			F
cetirizine	antihistamine	58	41 to 82	29	20 to 41	D
sumatriptan	antimigraine	50	35 to 71	1.5	1.1 to 2.1	J
glyburide	antidiabetic	48	12 to 190			A
alendronate	suppressant	46	33 to 66			J
methylprednisolone	glucocorticoid	44	13 to 150			D
clotrimazole	antifungal	44	26 to 77			J
oxycodone	opiod analgesic	43	25 to 74			J
estrone	hormone	39	14 to 110			J
bisoprolol	β-blocker	35	13 to 100	1.8	0.6 to 5.0	E
doxazosin	antihypertensive	32	8.1 to 130			J
clonazepam	antianxiety	32	8.8 to 120			F
amphetamine	CNS stimulant	30	6.2 to 150			J
dextroamphetamine	CNS stimulant	30	6.2 to 150			J
terazosin	antihypertensive	27	8.5 to 85			B
buspirone	antianxiety	24	5.7 to 100			F

(continued)

Table 1.1 (Continued)

Name	Classification	Predicted concentration excluding metabolism (ng/L)		Predicted concentration including metabolism (ng/L)		
		Geometric mean	Predicted range	Geometric mean	Predicted range	Excretion*
hydrocortisone	glucocorticoid	22	22			D
estradiol	hormone	22	2.5 to 190			J
diazepam	antianxiety	21	3.2 to 140			F
equilin	hormone	20	7.0 to 56			J
risperidone	antipsychotic	19	16 to 24			A
amlodipine	Ca channel blocker	14	6.8 to 27			B
felodipine	Ca channel blocker	13	6.7 to 27			B
lorazepam	antianxiety	12	1.9 to 69			C
alprazolam	antianxiety	11	1.9 to 69			F
17-α-dihydroequilin	hormone	11	4.0 to 33			J
norethindrone	hormone	11	11			F
ramipril	ACE inhibitor	10	3.5 to 28			F
neomycin	antibiotic	7.6	7.6			J
levothyroxine	hormone	5.2	2.9 to 9.4			J
glimepiride	antidiabetic	5.2	2.6 to 10			A
digoxin	cardiotonic	5.1	2.6 to 10	3.1	1.5 to 6.2	D
tobramycin	antibiotic	3.9	2.5 to 6.1			J
triamcinolone	glucocorticoid	3.5	3.5			D
mometasone	glucocorticoid	2.9	1.4 to 5.8			D
betamethasone	glucocorticoid	2.2	1.3 to 3.8			D
beclomethasone	glucocorticoid	2.1	2.1			J

(continued)

Table 1.1 (Continued)

Name	Classification	Predicted concentration excluding metabolism (ng/L)		Predicted concentration including metabolism (ng/L)		
		Geometric mean	Predicted range	Geometric mean	Predicted range	Excretion*
norgestimate	hormone	2.1	2.1			F
levonorgestrel	hormone	1.9	1.9			F
dexamethasone	glucocorticoid	1.3	0.82 to 2.0			J
ethinyl estradiol	hormone	1.2	1.2			J
fluticasone	antihistamine	1.2	0.19 to 7.6			J
timolol	β-blocker	1.2	0.58 to 2.3			J
clonidine	antihypertensive	0.90	0.37 to 2.2	0.40	0.16 to 1.0	D
nitroglycerin	antianginal	0.83	0.66 to 1.0			J
desogestrel	hormone	0.81	0.81			F
budesonide	glucocorticoid	0.70	0.70			J
tretinoin	keratolytic	0.67	0.12 to 3.7			J
latanoprost	antiglaucoma	0.60	0.60			J
salmeterol	bronchiodiolater	0.32	0.22 to 0.48			J
fluticasone	antiallergic	0.18	0.09 to 0.35			J
albuterol	bronchiodiolater	0.083	0.08			J

* Metabolism data obtained from Katzung (1998) and PDR (1999). Abbreviations indicate metabolic pathway: (A) extensive metabolism to inactive metabolites (B) extensive metabolism, possibly to conjugates; (C) excreted mostly as conjugates; (D) excreted mostly in original form (>50%); (E) excreted partially in original form (25-50%); (F) Extensive metabolism to active metabolites; (G) excreted as mixture of conjugates/original form; (H) excreted partially as conjugates (25-50%); (I) little excreted in urine; (J) data on metabolism not obtained.

It is instructivc to compare our estimates with estimates based on drug use data from Germany, where PhACs have been detected in wastewater (Table 1.2). For example, drugs that act as clofibric acid, such as etofibrate, precursors are extremely popular in Germany. However, they are rarely used in the United States because they have been replaced by HMG CoA reductase inhibitors. To illustrate difference between drug use in the US and Germany, we estimated concentrations of a group of PhACs in German wastewater by using the approach as described for the US. Results of these calculations indicate that drug use patterns vary considerably between the two countries. Expected concentrations are significantly higher in the US for fourteen of the PhACs while three of the compounds are expected at higher concentrations in Germany. The use of ten of the compounds varies by less than a factor of two between the US and Germany. Seven of the compounds in the list do not appear in industry survey data for the United States and cannot be compared with German estimates.

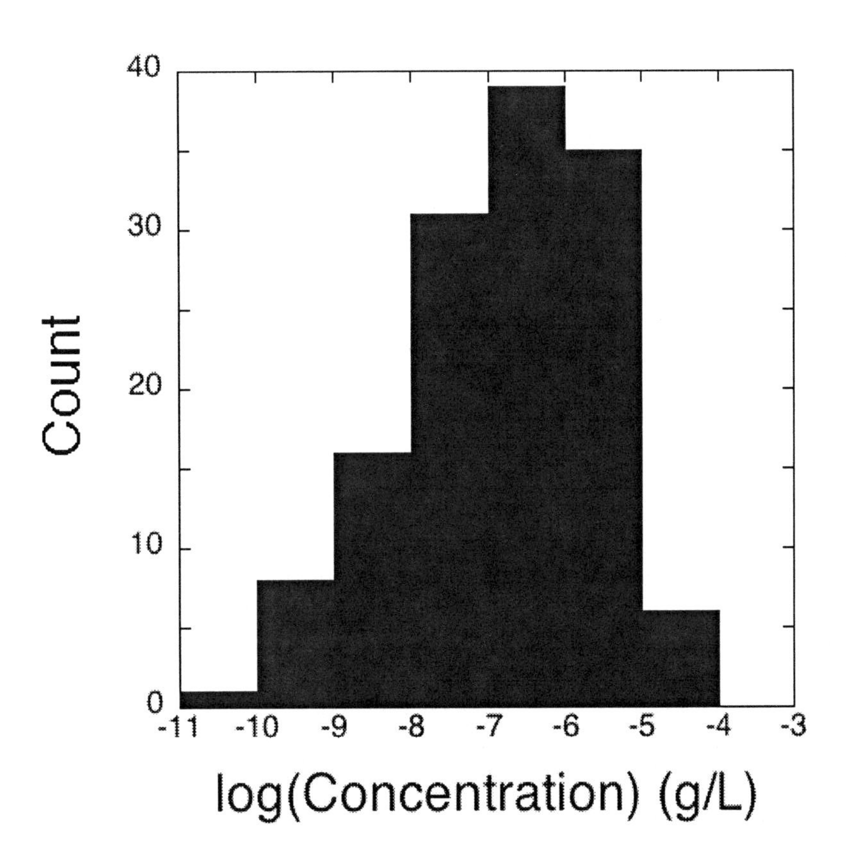

Figure 1.1 Histogram depicting the distribution of the geometric mean of the predicted wastewater concentrations for the compounds listed in Table 1.1

Table 1.2
Comparison of PhACs predicted in wastewater in the United States and Germany

Compound	Predicted concentration in wastewater (ng/L)*	
	Germany	United States
acetaminophen	8,200	62,000
acetylsalicyclic acid	28,000	a
amitriptyline	940	840
amoxicillin	7,300	27,000
atenolol	100	1,500
azithromycin	1,800	9,200
benzafibrate	4,900	b
betaxolol	53	c
bisoprolol	130	35
carbamazepine	9,200	c
cephalexin	67	14,000
ciprofloxacin	590	3,000
clarithromycin	210	2,800
diclofenac	3,600	c
diltiazem	1,600	2,600
erythromycin	2,200	1,500
fluoxetine	55	850
gemfibrozil	1,400	3,400
hydrochlorothiazide	610	1,900
ibuprofen	13,000	37,000
indometacine	660	c
ketoprofen	70	c
metoprolol	7,900	3,100
metformin	25,000	24,000
naproxen	150	2,500
paroxetine	30	440
penicillin	5,200	4,000
phenytoin	1,400	2,700
propranolol	690	250
ranitidine	3,800	3,000
sodium valproate	4,300	6,100
sulfamethoxazole	3,400	3,800
tramadol	1,500	2,200
trimethoprim	680	2,200

*Assumptions: Germany: population = 81.4 million; per capita wastewater production = 250 L per day; US: population = 250 million, per capita wastewater production = 320 L per day. Metabolism of drugs by patients and loss within the sewer system not considered. In addition, the estimates do not consider drugs obtained in hospitals and over the counter are not included. Abbreviations: a) over-the-counter medicine, not in top 200;b) Not in top 200, no longer commonly used in the U.S. c) Not in top 200. German prescription data obtained from Schwabe and Paffarth (1999).

Drugs Used in Animal Husbandry

Antibiotics are used in livestock both therapeutically to treat diseases and sub-therapeutically as feed additives to promote growth. Other PhACs also are used in animal husbandry (e.g., growth hormones) but normally at much lower doses than the antibiotics. In our assessment we assumed that other drugs used in animal husbandry are insignificant compared to the antibiotics. Quantification of antibiotic use in livestock and aquaculture is challenging because drug use is often not documented and some drugs can be purchased from distributors of animal feed without reporting requirements. Furthermore, the formulae of antibiotics employed as feed additives often are not revealed by the feed manufacturers.

Because published data on the amount of antibiotics used in agriculture are not available, predictions were made based upon annual animal feed consumption and the recommended doses of antibiotics added in feed. Although antibiotics are also given to animals to treat disease, it is difficult to estimate the frequency and quantities of antibiotics use for these purposes. It is assumed that antibiotics used as feed additives to promote animal weight gain and feed efficiency account for the majority of antibiotic consumption in livestock because of their continuing usage. For instance, it was estimated that 8.2 million kilograms of antibiotics were used in major species of food animals in 1985; 90% of which was used for sub-therapeutic dose application.

Antibiotics are also used in aquaculture. In the United States, aquaculture production is relatively small compared to livestock production and is concentrated in the coasts and estuaries of a few states such as Washington and Mississippi. Antibiotic use in aquaculture could result in localized water pollution. Because of the relatively small quantities of use and localized contamination, antibiotics used in aquaculture are not included in our estimation.

Our predictions are based on an estimate of the mass of each antibiotic consumed in promoting livestock growth, which is converted into concentrations of antibiotics in the liquid waste generated by animal feeding operations (AFOs). The prediction methods are described in the following paragraphs.

The mass of each antibiotic used to promote animal growth is calculated in two ways: (1) the annual consumption of each antibiotic per animal; and, (2) the annual consumption of each antibiotic by all animal species. The first approach provides information needed to calculate antibiotic loading in raw liquid animal waste at a feeding operation. The second approach provides information on the total consumption for each antibiotic.

The annual consumption of an antibiotic per animal was calculated by multiplying the quantities of feed consumed per animal by the concentration of antibiotic in feed. The amount of feed consumed per animal per year can be calculated by multiplying the grain used per "grain consuming animal unit" (1.87×10^3 kg/year, Feed Yearbook, USDA, 2000) by the equivalency factors for animal species. The "animal unit" (AU) is a unit of measurement used to standardize sizes of animal feeding operations (AFOs). The number of AUs is determined by multiplying the number of animals of each species (other than poultry) by an equivalency factor. Species equivalency factors are 1.0 for slaughter/feeder cattle, 1.4 for mature dairy cattle and 0.4 for swine (>55lbs).

The amount of feed consumed by poultry (broilers and layers) was calculated in a different way. (A broiler is raised for consumption while a layer is raised for egg production.) A broiler has a 6-7 week lifespan and consumes approximately 8 lbs (i.e., 3.6 kg) of feed during its lifespan. In addition, there is a typical 2-week downtime between two crops of broilers in poultry operations. It was assumed that there are 6.5 crops of broilers per year at a poultry operation. This leads to approximately 52 lbs (i.e., 23.6 kg) of feed consumption per year per broiler space. Information was also gathered on the amount of feed consumed by a layer. A layer has a typical 65-week lifespan and consumes approximately 0.23 lbs (i.e., 0.10 kg) of feed per day. This figure can be converted to a feed consumption rate of 38.1 kg/year-layer.

The amount of feed consumed per animal was then multiplied by the concentration of antibiotic in feed to obtain the mass of each antibiotic consumed per animal per year. Because information on the feed formulation and antibiotic additives are not publicly accessible and are difficult to obtain, the recommended dosages of antibiotics listed in the Feed Additive Compendium (2000) were used. The Feed Additive Compendium is updated on a yearly basis and information for year 2000 was used for calculation. The calculated results are listed in Table 1.3. When a range of dosages is recommended, calculations were performed with the minimum and maximum dosages, respectively.

Table 1.3
Antibiotic (in feed) consumption per year per animal (g/year-animal)

Antibiotic	Category	Broiler		Layer		Cattle		Swine	
		min.	max.	min.	max.	min.	max.	min.	max.
Amprаymycin	aminoglycoside	NA*		NA		NA		33.7	67.4
Arsanilic Acid	other	2.1		2.1		NA		33.7	67.4
Bacitracin MD†	polypeptide	0.1	1.2	0.1	1.2	NA		33.7	67.4
Bacitracin Zinc	polypeptide	0.1	1.2	0.2	0.6	65.5	131.0	15.0	30.0
Bambermycins	aminoglycoside	<0.1	<0.1	NA		1.9	74.9	1.5	3.0
Carbadox	other	NA		NA		NA		7.5	18.7
Chlorotetracycline	tetracycline	0.2	1.2	0.2	1.2	131.0		7.5	18.7
Lincomycin	aminoglycoside	<0.1	0.1	<0.1	<0.1	NA		15.0	
Oxytetracycline	tetracycline	1.2	2.4	1.2	2.4	140.4		7.5	37.4
Penicillin	β-lactam	0.1	1.2	<0.1	1.2	NA		7.5	37.4
Roxarsone	other	0.5	1.1	0.5	1.1	NA		17.0	25.5
Tiamulin	other	NA		NA		NA		7.5	
Sulfamethazine	sulfonamide	NA		NA		NA		37.4	
Tylosin	macrolide	0.1	1.2	0.1	1.2	NA		37.4	
Virginiamycin	macrolide	0.1		<0.1		20.6	30.0	3.7	7.5

*Not applicable
† Bacitracin methylene disalicylate.

To calculate the masses of antibiotics consumed by all animals, information on the total annual feed consumption by all livestock in the United States was needed. The total feed consumption in year 2000 was calculated by multiplying the number of "grain consuming animal units" (30.1, 23.0 and 21.1 million units for poultry, swine and cattle respectively) by the amount of grain consumed per "grain consuming animal unit" (i.e., 1.87×10^3 kg) (Feed Yearbook, USDA, 2000). The mass of antibiotics consumed was then calculated by multiplying the total mass of feed consumed by the recommended dosages of antibiotics.

It was also necessary to consider the fact that only one or two of the recommended antibiotics are used in feed at any time and the selection of antibiotics depends on the feed manufacturer. Information regarding the selection of antibiotics is difficult to obtain, thus we assumed that each antibiotic was added in approximately 50% of the consumed feed. The results are listed in Table 1.4. Combining the amounts of antibiotics consumed by all animal species yields the total mass of consumption for each antibiotic. These calculations are only rough approximations for the quantities of growth-promoting antibiotics; however they allow us to identify "high-use" veterinary antibiotics.

As shown in Table 1.3 and 1.4, there are 14 antibiotics commonly used for promoting livestock growth. Considerable differences in antibiotics exist among different animal species. For example, only four antibiotics are used in cattle while fourteen are used in swine. Among the 14 antibiotics, bacitracin, oxytetracycline, chlorotetracycline, bambermycin and virginiamycin were used among all animal species. According to our estimate, bacitracin, oxytetracycline, and arsanilic acid are the top three most common growth-promoting antibiotics.

There were approximately 450,000 animal feeding operations (AFOs) throughout the United States, ranging from small livestock production facilities with a few animals to the large and geographically concentrated facilities generating a mass of animal waste equivalent to the waste produced by humans in a medium-sized city. Under Section 502 of the Clean Water Act, confined animal feeding operations (CAFOs) are point sources and must apply for a National Pollutant Discharge Elimination System (NPDES) permit. Of approximately 6,600 CAFOs, fewer than a quarter have NPDES permits (EPA, 1996). CAFO liquid waste usually undergoes some type of pretreatment prior to land application. In some cases, it is combined with other wastewater for further treatment before discharge.

Table 1.4
Antibiotic consumption rates (kg/year) for poultry, cattle and swine in the United States

Antibiotic	Category	Poultry		Cattle		Swine		Total	
		min.	max.	min.	max.	min.	max.	min.	max.
Bacitracin*	Polypeptide	225	2817	711	1422	1399	2799	2336	7038
Oxytetracycline	Tetracycline	1409	2817	1523		215	1076	3147	5417
Arsanilic Acid	Other	2536	2536	NA†		969	1938	3504	4473
Chlorotetracycline	tetracycline	282	1409	1422		215	538	1919	3369
Penicillin	β-lactam	56	1409	NA		215	1076	272	2485
Tylosin	macrolide	113	1409	NA		1076		1189	2485
Roxarsone	Other	640	1279	NA		489	732	1128	2011
Ampraymycin	aminoglycoside	NA		NA		969	1938	969	1938
Sulfamethazine	sulfonamide	NA		NA		1076		1076	
Bambermycin	aminoglycoside	28	56	20	812	43	86	92	955
Virginiamycin	macrolide	6	141	223	325	108	215	337	681
Lincomycin	aminoglycoside	56	113	NA		431		487	543
Carbadox	Other	NA		NA		215	538	215	538
Tiamulin	Other	NA		NA		215		215	

* Bacitracin methylene disalicylate
†.Not applicable.

The wastewater volume from CAFOs consists of: (i) waste quantities being generated by animals (also called pollutant load); and, (ii) water added to the waste from sources such as flushwater to remove manure from alleys and barns, water for cleaning, rainfall runoff from roofs and open lots and direct rainfall on pretreatment facilities (Overcash et al., 1983).

Water use varies considerably from one operation to another, depending on such factors as type of buildings, methods of flushing, and type of management. In general, Overcash et al. (1983) suggest that the volume of flushwater used in swine and poultry facilities can be estimated by calculating approximately 2 gallons per minute (gpm) of water per 100 pounds (lbs) of animal weight for the flushing period. For cattle and dairy facilities, 40 to 50 gallons per cow per day are assumed in flushing requirements for freestall alleys. The frequency of daily flushing will determine the total volume of flushwater used.

Using the above information and assuming an average poultry weight of 8 lbs and an average swine weight of 60 lbs, the flushwater flows computed for CAFOs (2500 head) of poultry and swine are approximately 400 and 3000 gallons per minute respectively. It was also assumed that one flushing period of 30 minutes is employed per day; this leads to 4.54×10^4 and 3.41×10^5 L/day of flushwater for poultry and swine CAFOs respectively. The volumes of animal wastes can also be estimated based upon species populations (The Agricultural Waste Management Field Handbook, USDA, 1992). It was found that animal waste volume was insignificant compared to the flushwater volume, thus flushwater volume roughly represents the total volume of the CAFO wastewater. For a typical cattle CAFO (2500 head), 50 gallons of flushwater was assumed per cow per day, leading to a flushwater flow of 4.73×10^5 L/day.

The estimated concentrations of antibiotics in raw CAFO wastewater were obtained by dividing the amounts of antibiotics consumed per animal per day (Table 1.3) by the wastewater volume generated per animal per day. The calculated results are listed in Table 1.5. The calculations also assume 80% of antibiotics were excreted without undergoing metabolism. The estimated concentrations ranged from 0.2 μg/L to greater than 1.6 mg/L. There are considerable differences in antibiotic use among different animal species.

Comparing Tables 1.1 and Table 1.5, it is evident that antibiotics used in livestock are different from the ones used in human therapy. This approach has been adopted to reduce the risks of development of resistant bacteria in animals, which may in turn be passed on to humans, thus diminishing the effectiveness of antibiotics in treatment of human diseases. A comparison between human and agricultural uses of antibiotics is summarized in Table 1.6. The table indicates some important differences among the different antibiotics. For instance, tetracyclines are much more important in agricultural applications than for human therapy. Barcitracin, tylosin and virginiamycin are used almost exclusively for agricultural purposes. In general, different compounds of β-lactams, macrolides, fluoroquinolones, and sulfonamides are used separately for humans and for food animals. For instance, fluoroquinolones (e.g., ciprofloxacin) play an important role in treating human diseases; many of the new antibiotics being used to fight antibiotic-resistant bacteria are members of fluoroquinolones. Other fluoroquinolones such as enrofloxacin and sarafloxacin were developed to prevent and cure diseases in poultry. However, due to the concern of fluoroquinolone residues in meat products, USDA has recently issued a ban on the use of enrofloxacin and sarafloxacin in poultry industry (C&EN, 2000). For the early-developed sulfonamide antibiotics, only sulfamethazine is used as a feed additive. A variety of sulfonamide antibiotics are used in livestock for disease treatment, but at much lower application rates.

Table 1.5
Estimated concentrations of antibiotics in CAFO wastewater (μg/L)

Antibiotic	Broiler		Layer		Cattle		Swine	
	min.	max.	min.	Max.	Min.	max.	min.	max.
Amprаymycin	NA*		NA		NA		542	1,084
Arsanilic Acid	256		256		NA		542	1,084
Bacitracin MD[†]	11	142	11	142	NA		542	1,084
Bacitracin Zinc	11	142	28	71	759	1,517	241	482
Bambermycins	3	6	0.2		22	867	24	48
Carbadox	NA		NA		NA		120	301
Chlorotetracycline	28	142	28	142	1,517		120	301
Lincomycin	6	11	0.5	0.5	NA		241	241
Oxytetracycline	142	285	142	285	1,626		120	602
Penicillin	7	142	7	142	NA		120	602
Roxarsone	65	129	65	129	NA		273	409
Tiamulin	NA		NA		NA		120	
Sulfamethazine	NA		NA		NA		602	
Tylosin	11	142	11	142	NA		602	
Virginiamycin	14		0.6		238	347	60	120

* Not applicable
[†].Bacitracin methylene disalicylate.

Table 1.6
Antibiotics currently used in human therapy and agriculture

	Application				Application		
Compound	Human Therapy*	Agric. Feed	Agric. Ailment	Compound	Human Therapy[a]	Agric. Feed	Agric. Ailment
β-lactam:				*aminoglycoside:*			
amoxicillin	X		X	neomycin	X		X
cephalexin	X			tobramycin	X		
penicillin	X	X	X	apramycin		X	X
cefprozil	X			bambermycin		X	X
cefuroxime	X			lincomycin		X	X
loracarbef	X			efrotomycin			X
ampicillin	X"		X	gentamycin			X
				streptomycin			X
macrolide:				*sulfonamide:*			
azithromycin	X			sulfamethoxazole	X		
clarithromycin	X			sulfamethazine		X	X
erythromycin	X		X	sulfachloropyridazine			X
oleandomycin			X	sulfadimethoxine			X
roxithromycin	X"			sulfaethoxypyridazine			X
spectinomycin			X	sulfamerazine			X
tilmicosin			X	sulfathiazole			X
tylosin		X	X	sulfamethiazole			X
virginiamycin		X	X				
fluoroquinolone:				*tetracycline:*			
ciprofloxacin	X			tetracycline	X		X
levofloxacin	X			doxycycline	X"		X
norfloxacin	X"			chlortetracycline		X	X
enrofloxacin			X	oxytetracycline		X	X
sarafloxacin			X				
β-lactamase inhibitor:				*Other:*			
clavulanic acid	X			trimethoprim	X		
				mupirocin	X		
				barcitracin MD*		X	X

(continued)

Table 1.6 (Continued)

Compound	Application: Human Therapy[a]	Agric. Feed	Agric. Ailment	Compound	Application: Human Therapy[a]	Agric. Feed	Agric. Ailment
				barcitracin zinc		X	X
				arsanilic acid		X	X
				carbodox		X	X
				roxarsone		X	X
				ivermectin			X

* X = ranks among the top 200 prescription drugs; X" = used for human therapy but does not rank below the top 200 prescription drugs

†.Bacitracin methylene disalicylate.

OCCURRENCE DATA

Another important tool for identifying candidate compounds is occurrence data from other scientific studies. At the time that this project started, published data on PhACs in municipal wastewater effluents and surface waters was limited to studies conducted in Germany and Switzerland. However, results from studies conducted in Canada and the United States were reported after the compounds to be analyzed were selected. To identify compounds to be analyzed during the occurrence survey, published data were reviewed to identify compounds that could be analyzed readily and the less common compounds that still might be detectable in wastewater effluent. In addition, the occurrence data provide guidance on the removal of compounds during municipal wastewater treatment.

As mentioned above, most data of prescription drugs and over-the-counter medications in municipal wastewater have been collected by German and Swiss researchers (Table 1.7). The most comprehensive study of PhACs in municipal wastewater was published by Ternes (1998). In this study, a total of 32 PhACs were measured in wastewater effluent samples collected at treatment plants throughout Germany. The most comprehensive database for detection of PhACs in surface waters is the study by Kolpin et al. (2002), of the US Geological Survey (USGS) who studied the occurrence of PhACs and wastewater tracers in surface waters throughout the US (Table 1.7). The frequency of occurrence in the study by Kolpin et al. (2002) was somewhat less than that of the German and Swiss researchers. This discrepancy was caused mainly because the USGS studied rivers where dilution of wastewater effluent was important whereas the other researchers focused on effluent-dominated waters. Some of the compounds reported in Table 1.7 do not appear in the list of popular US prescription drugs (i.e., Table 1.1). Compounds detected with a high frequency in previous studies (i.e., in more than 50% of the wastewater effluents sampled) included:

- Benzafibrate, other fibrate-based drugs and their metabolites (clofibric acid, fenofibric acid). [As mentioned previously, these compounds are no longer popular in the U.S. and are not expected to be present at significant concentrations in US wastewater.]
- Analgesics (diclofenac, indometacine, ketoprofen, phenazone), many of which also are available in over-the-counter formulations.
- The beta-blocker, nadolol.
- The antiepiletic, carbamazepine, which also is less widely prescribed in the US, relative to Germany, where it has been detected frequently.

Table 1.7
Summary of occurrence data for prescription drugs excluding antibiotics

	Wastewater Effluent*			Surface Water*			Drinking Water*		
Compound	ND	< 50%	>50%	ND	< 50%	>50%	ND	< 50%	>50%
acetaminophen		1		1					
acetylsalicylic acid		1	5,6,7	7		1	7		
betaxolol			1,4		1	4	4		
bezafibrate			1,5,7	5		1,7		7	
bisoprolol			1,4		1	4	4		
carazolol		1,4			1,4		4		
carbamazepine		8	1			1			
cimetidine					10				
clenbuterol		1,4		4	1		4		
clofibrate	1			1					6
clofibric acid			1,5,7		5	1,2,7			2,3,7
codeine					10				
cyclophosphamide		1		1					
diazepam		1	6	1		6			
diclofenac			1,5,7-9			1,7		7	
digoxin				10					
diltiazem					10				
dimethylaminophenazone		1			1				
enalaprilat					10				
etofibrate	1			1					
fenofibrate		1		1					
fenofibric acid			1,5,7			1,7	7		
fenoprofen	1,7			1,7			7		
fenoterol		1,4			1,4		4		
fluoxetine					10				
gemfibrozil			1,5,7,8		10	1,7	7		
gentisic acid		1			1				
ibuprofen			1,5,7			1,5,7		7	
ifosfamide		1		1					
indometacine			1,5,7			1,7	7		
ketoprofen			1,5,7	7	1		7		
meclofenamic acid	1			1					
metoprolol			1,4			1,4	4		
metformin					10				
nadolol			1,4	1	4		4		
naproxen			1,5			1,5			
o-hydroxyhippuric acid	1			1					

(continued)

Table 1.7 (Continued)

	Wastewater Effluent*			Surface Water*		
Compound	ND	< 50%	>50%	ND	< 50%	>50%
phenazone			1			1
propranolol			1,4			1,4
rantidine					10	
salbutamol		1,4		4,10	1	
salicylic acid		1				1
terbutalin		1,4			1,4	
timolol		1,4			1	4
tolfenamic acid	1			1		
warfarin				10		

*Indicates frequency with which compound was detected in samples from each medium. ND = compound never detected; >50%=compound detected in more than 50% of samples analyzed; <50%=compound detected in less than 50% of samples analyzed References are as follows: (1)Ternes (1998); (2) Stan, Heberer, and Linkerhagner (1994); (3) Heberer and Stan (1996a,b); (4) Hirsch et al. (1998); (5) Stumpf et al. (1999); (6) Richardson and Bowron (1985);(7) Stumpf et al. (1996); (8) Drewes JE, Heberer T, Reddersen K (2002); (9) Buser HR, Poiger T, Muller MD (1998); (10) Kolpin et al. (2002)

Occurrence data for antibiotics are reviewed and summarized in Table 1.8. Most previous studies were conducted in Europe where antibiotic use could be different from that of the United States. However, the previous studies provide guidance for identifying the classes of antibiotics that are more persistent in the environment.

In general, β-lactam antibiotics were not detected in most environmental waters. The β-lactam compounds are readily hydrolyzed in the environment (Hou and Poole, 1969), and thus are less likely to be persistent. Other classes such as fluoroquinolones, macrolides, sulfonamides and tetracyclines have been detected. Occurrence data are not available for aminoglycoside antibiotics and most of the other types of antibiotics.

After reviewing data on the use patterns of prescription drugs, occurrence data and the availability of analytical techniques, several groups of PhACs were identified for use in the occurrence survey. The selection method involved consideration of analytical instruments available in the participating laboratories and the likelihood of detecting the PhAC. In some cases, the analytical method selected would allow for simultaneous analysis of a PhAC that was not expected to be present at high concentrations. In these situations, the compound was added the analyte list despite the fact that it was not expected to be present at elevated concentrations.

The first group of PhACs selected was the acidic drugs, including diclofenac, gemfibrozil, ibuprofen and naproxen. The acidic PhACs indometacine, nadolol and ketoprofen also were included in method development because they could be analyzed using the same methods. At the time the project began, all of the abovementioned compounds had been detected in wastewater effluent samples in Europe and the prescription data (Table 1.1) suggested that the gemfibrozil, ibuprofen and naproxen would be prevalent in wastewater at concentrations above 1000 ng/L.

The second group of PhACs selected was the beta-blockers. The compound metoprolol had been detected in wastewater effluent in Germany and prescription data indicated that it would be present in wastewater at a concentration above 1000 ng/L. The beta-blocker propranolol also was expected to be present in wastewater, but concentrations in the US were expected to be less than 1000 ng/L.

Among the antibiotics, the selection of compounds to be analyzed was based upon compounds expected to be present in wastewater and those used for agricultural purposes. For wastewater, the compounds ciprofloxacin and sulfamethoxazole and trimethoprim were chosen on the basis of their expected concentrations in wastewater and occurrence data from Europe. Among the antibiotics used for animal husbandry, selfamethazine was chosen based on expected use and persistence in the environment. After selecting these compounds, it became evident that the analytical methods used to analyze these compounds also would allow us to detect the fluoroquinolones enrofloxacin, norfloxacin and ofloxacin.

Table 1.8

Summary of occurrence data for antibiotics

	Wastewater Effluent			Surface Water			Ground Water		
Compound	N.D.	<50%	>50%	N.D.	<50%	>50%	N.D.	<50%	>50%
β-lactams:									
cloxacillin	2			2			2		
dicloxacillin	2			2			2		
methicillin	2			2			2		
nafcillin	2			2			2		
oxacillin	2			2			2		
penicillin g	2			2			2		
penicillin v	2			2			2		
macrolides:									
clarithromycin			2, 5		2		2		
erythromycin-h_2o			2, 5		6	2	2		
roxithromycin			2, 5		2,6		2		
quinolones:									
ciprofloxacin			3		6		2		
sulfonamides:									
sulfadiazine			1		1				
sulfamethazine	2, 5			2	6		2		
sulfamethizole			1	1	6				
sulfamethoxazole			1,2,5		6	1, 2	2		
tetracyclines:			4		4				
chlorotetracycline	2			2	6		2		
doxycycline	2			2,6			2		
oxytetracycline	2			2	6		2		
tetracycline	2			2	6		2		
others:									
chloramphenicol	5	2			2		2		
trimethoprim			2, 5		2,6		2		

*Indicates frequency with which compound was detected in samples from each medium. ND = compound never detected; >50%=compound detected in more than 50% of samples analyzed; <50%=compound detected in less than 50% of samples analyzed References are as follows: (1) Hartig et al., (1999); (2) Hirsch et al.,(1999); (3) Hartmann et al.(1998); (4) Meyer et al. (2000) ; (5) McArdell et al., (2003) (6) Kolpin et al. (2002).

CHAPTER 2
ANALYTICAL METHOD DEVELOPMENT AND TESTING

DRUGS USED FOR HUMAN THERAPY (EXCLUDING ANTIBIOTICS)

The analysis of acidic drugs and beta-blockers was performed by gas chromatography/tandem mass spectrometry (GC/MS/MS) after solid phase extraction (SPE) and derivatization. Due to the different functional groups on these two classes of PhACs, two different analytical methods were used. For the acidic drugs, the method described by Stumpf et al. (1996) was used as a guide in method development and Ternes et al. (1998) was used for the beta-blockers. As a result of the numerous steps in the analysis and the complex nature of the extracts being analyzed, it was difficult to obtain accurate and precise results, especially when analyzing wastewater effluent samples. As a result, considerable attention was focused on identifying steps in which errors could occur and developing methods that were as robust and reproducible as possible. The following sections describe the final analytical methods adopted for the occurrence survey and the method development and validation studies performed in conjunction with these analyses.

Analytical Methods for Acidic Drugs

Grab samples were collected in 1-L glass bottles with Teflon-lined screw caps. Each bottle was kept in an individual polyethylene bag. Prior to sampling, bottles were cleaned with Micro brand laboratory detergent, rinsed with water followed by HPLC grade methanol and deionized water between each analysis. Bottles were shipped to participating utilities in coolers with blue ice packs and were returned by overnight mail within one day of sample collection.

For samples collected from wastewater treatment plants or water treatment plants using chlorine for disinfection 0.5 g/L of $Na_2S_2O_3$ was added to the samples bottle as a preservative. Samples from other locations were collected in bottles without any added preservative. Each set of samples was shipped with a field blank, consisting of deionized water, which was analyzed with the samples. PhACs were never detected in the field blanks. Samples were collected by field personnel who were familiar with trace organic sampling protocols. Field personnel wore polyethylene gloves when handling bottles and were instructed to minimize the amount of time that the bottle is kept uncapped outside of the cooler. Upon arrival at UC Berkeley, samples and log sheets were visually inspected and transferred to a 5°C storage area. Samples were extracted as soon as practical and within no more than 72 hours after arrival.

Each set of ten samples was extracted in a batch along with appropriate QA/QC standards. The following samples were included with each set of samples:

(1) Field blank (1 L of deionized water that traveled to and from the field site);
(2) Matrix recovery sample (1 sample from the site spiked with all analytes at 1,000 ng/L);
(3) Duplicate sample;
(4) Auxiliary standard consisting of a mixture of the derivatized analytes, as prepared by a third party in the laboratory.

Each sample was filtered through 0.45-μm glass fiber filters prior to solid phase extraction. Following filtration, mecoprop-^{13}C (CDN isotopes, Quebec, Canada) was added as a surrogate standard at a final concentration of 1000 ng/L. For samples that contained relatively high concentrations of organic matter (e.g., wastewater effluent, water from engineered treatment wetlands) a sample volume of 0.5 to 1 L was used. For samples that contained less organic matter (e.g., reverse osmosis permeate, deep groundwater) a sample volume of 1 to 2 L was used. For the acidic drugs, the sample pH was adjusted to pH less than 2 with concentrated sulfuric acid. The sample was placed in a silanized glass container connected to the extraction columns with Teflon tubing. The silanized glass extraction columns were packed with 500 mg of ENVI-18 solid phase extraction resin (Supelco). The samples were passed through the extraction columns at a flow rate of approximately 20 mL/min by connecting the tubing to a peristaltic pump placed downstream of the extraction column. After passing the sample through the columns, they were dried for 10 minutes by pumping air through the resins at the same flow rate.

Following solid phase extraction, the samples were eluted from the resins using 10 mL of HPLC grade methanol (Fisher Scientific). The methanol was used to rinse the sample container prior to pumping it through the extraction column at the same flow rate as the samples. The methanolic extracts were collected in silanized glass test tubes. The extracts then were dried completely by placing them in a vacuum oven at room temperature overnight. Following the drying step, the samples were resuspended in 2 mL of HPLC grade methanol and transferred to 5 mL glass vials. The extracts then were blown to dryness under a gentle stream of high purity nitrogen over a period of approximately 30 minutes. Following blowdown, the samples were derivatized with a diazomethane/diethylether mixture, prepared as described by the manufacturer (Aldrich Chemical Company, Technical Bulletin A1-180). Diazomethane presents some potential safety hazards (e.g., it is a mutagen and can explode upon impact) and should be handled with great care (see Aldrich Chemical Company, Technical Bulletin A1-180 for details). After adding 250 μL of diazomethane/diethyl ether, the extracts were allowed to react for 2 minutes prior to quenching the excess diazomethane with 10 mL of a 1:10 acetic acid/acetone mixture. The derivatized samples were again blown to near dryness under a stream of high purity nitrogen and resuspended in 250 to 1000 μL of isooctane with 1000 mg/L of hexachlorobenzene as an internal standard.

The derivatized acidic drugs were analyzed by GC/MS/MS using a Finnegan GCQ GC/MS/MS system with a 30-meter DB-5 column. Prior to analysis the GC/MS/MS system was optimized by changing the injection port liner, cutting back the first 2-cm of the column when necessary, and cleaning the ion trap, when necessary. The analysis of wastewater extracts led to a decrease in system performance, and it was usually necessary to repeat the cleaning procedure after approximately 200 injections. The analytical conditions used for quantification are listed in Table 2.1 and the following conditions were used for the oven and mass spectrometer: isothermal at 50°C for 4 min, 20 °C/min to 120°C, 2°C/min to 180 °C, 30°C/min to 290 °C, where it was held for 8 min. The carrier gas used was helium at a flow rate of 1.2 mL/min. Splitless injection of 2 μL samples was used with a split flow of 50 mL/min and an injection port temperature of 270 °C. The mass spectrometer had a source temp of 200 °C and the transfer line was held at 300 °C.

Table 2.1
Conditions used for analysis of acidic drugs

Compound	Retention time (min)	Parent ion	Product ion	Voltage (V)
Ibuprofen	19.2	161	105, 119	0.75
Mecoprop-C_{13}	19.6	231	145,172	1.0
Gemfibrozil	32.8	143	83	0.75
Naproxen	41.0	244	170, 185	0.75
Ketoprofen	46.3	209	105	0.50
Diclofenac	48.7	214	151, 178	0.75
Indometacine	58.5	371	139, 312	1.0
Hexachlorobenzene	25.8	142, 249, 284*		

*Hexacohlorbenzene, the internal standard, was quantified by SIM.

Prior to analysis of samples two standard curves were constructed using a mixed standard of the suite of acidic drugs. The low concentration standard curve contained 37.5 μg/L, 75 μg/L, 150 μg/L, 225 μg/L and 300 μg/L of the compounds (expressed in terms of the concentration in the pre-concentrated sample extracts) while the high concentration standard curve used 300 μg/L, 600 μg/L, 900 μg/L and 1200 μg/L. A linear calibration curve was calculated from simple linear regression. Following calibration, a run sequence was used consisting of five standards followed by a randomized mixture of the samples and QA/QC samples. The calibration curve was checked every ten samples by running a blank and a reslope standard from the middle of the calibration curve. If the calibration standard disagreed with the standard curve by more than 25% the samples in the following section were rerun.

For the acidic drugs, the target for recoveries was 60-120%, as determined by recovery of C_{13} mecoprop. In cases in which duplicate samples were run, and only one duplicate met the criterion, the other value was not used. If neither value met the criterion, the data were reported with appropriate qualifiers. Typical detection limits for acidic drugs in wastewater effluent were around 10 ng/L.

Analytical Methods for Beta-Blockers

Grab samples were collected in 1-L glass bottles with Teflon-lined screw caps. Each bottle was kept in an individual polyethylene bag. Prior to sampling, bottles were cleaned with Micro brand laboratory detergent, rinsed with water followed by methanol and deionized water between each analysis. Bottles were shipped to participating utilities in coolers with blue ice packs and were returned by overnight mail within one day of sample collection.

For samples collected from wastewater treatment plants or water treatment plants using chlorine for disinfection, 0.5 g/L of $Na_2S_2O_3$ was added to the samples bottle as a preservative. Samples from other locations were collected in bottles without any added preservative. Each set of samples was shipped with a field blank, consisting of deionized water, which was analyzed with the samples. Samples were collected by field personnel who were familiar with trace organic sampling protocols. Field personnel wore polyethylene gloves when handling bottles and were instructed to minimize the amount of time that the bottle is kept uncapped outside of the cooler. Upon arrival at UC Berkeley, samples and log sheets were visually inspected and transferred to a 5°C storage area. Samples were extracted as soon as practical and within no more than 72 hours after arrival.

Each set of ten samples was extracted in a batch along with appropriate QA/QC standards. The following samples were included with each set of samples:

(1) Field blank (1 L of deionized water that traveled to and from the field site);
(2) Matrix recovery sample (1 sample from the site spiked with all analytes at 1,000 ng/L);
(3) Duplicate sample;
(4) Auxiliary standard consisting of a mixture of the derivatized analytes, as prepared by a third party in the laboratory.

Each sample was filtered through 0.45-μm glass fiber filters prior to solid phase extraction. For samples that contained relatively high concentrations of organic matter (e.g., wastewater effluent, water from engineered treatment wetlands a sample volume of 0.5 to 1 L was used. For samples that contained less organic matter (e.g., reverse osmosis permeate, deep groundwater) a sample volume of 1 to 2 L was used. The sample was placed in a silanized glass container connected to the extraction columns with Teflon tubing. The silanized glass extraction columns were packed with 500 mg of ENVI-18 solid phase extraction resin (Supelco). The samples were passed through the extraction columns at a flow rate of approximately 20 mL/min by connecting the tubing to a peristaltic pump placed downstream of the extraction column. After passing the sample through the columns, they were dried for 10 minutes by pumping air through the resins at the same flow rate.

Following solid phase extraction, the samples were eluted from the resins using 10 mL of HPLC grade methanol (Fisher Scientific). The methanol was used to rinse the sample container prior to pumping it through the extraction column at the same flow rate as the samples. The methanolic extracts were collected in silanized glass test tubes. The extracts then were dried completely by placing them in a vacuum oven at room temperature overnight. Following the drying step, the samples were resuspended in 2 mL of HPLC grade methanol and transferred to 2 mL volumetric flasks. The extracts then were blown to dryness under a gentle stream of high purity nitrogen over a period of approximately 30 minutes. Following blowdown, the samples were derivatized with a 50 μL of N-methyl-N-(trimethylsilyl)trifluoroacetamide (MSTFA) from Sigma Chemical Company at room temperature for 30 minutes. The extracts then were sealed by placing Teflon tape and Parafilm over the ground glass toppers. The volumetric flasks were heated to 60°C for 10 minutes and 10 μL of N-methyl-bis(trifluoroacetamide) (MBTFA) was added. The samples were cooled and resuspended in 250 to 1000 μL of isooctane with 1000 mg/L hexachlorobenzene as an internal standard. It should be noted that this internal standard does not account for variations in the efficiency of derivatization.

The derivatized beta-blockers were analyzed by GC/MS/MS using a Finnegan model GCQ GC/MS/MS system with a 30-meter DB-5 column. Prior to analysis the GC/MS/MS system was optimized by changing the injection port liner, cutting back the first 2-cm of the column when necessary, and cleaning the ion trap, when necessary. The analysis of wastewater extracts led to a decrease in system performance, and it was usually necessary to repeat the cleaning procedure after approximately 200 injections. The analytical conditions used for quantification are listed in Table 2.2 and the following conditions were used for the oven and mass spectrometer: isothermal at 50°C for 4 min, 16°C/min to 180°C, 5°C/min to 250°C. The carrier gas used was helium at a flow rate of 1.2 mL/min. Splitless injection of 2 μL samples was used with a split flow of 50 mL/min and an injection port temperature of 230 °C. The mass spectrometer had a source temp of 200 °C and the transfer line was held at 250°C.

Prior to analysis of samples two standard curves were constructed using a mixed standard of the suite of beta-blockers. The low concentration standard curve contained 37.5 μg/L, 75 μg/L, 150 μg/L, 225 μg/L and 300 μg/L of the compounds (expressed in terms of the concentration in the pre-concentrated extracts) while the high concentration standard curve used 300 μg/L, 600 μg/L, 900 μg/L and 1200 μg/L. A linear calibration curve was calculated from simple linear regression. Following calibration, a run sequence was used consisting of five standards followed by a randomized mixture of the samples and QA/QC samples. The calibration curve was checked every ten samples by running a blank and a reslope standard from the middle of the calibration curve. If the calibration standard disagreed with the standard curve by more than 25% the samples in the following section were rerun.

For the beta-blockers, the target for recoveries were lower than that obtained for the acidic drugs. A target value of 60-120%, as determined by spike recovery samples in which 1000 ng/L of a mixed standard was added to a sample included in the sample set. In cases in which duplicate samples were run, and only one duplicate met the criterion, the other value was not used. If neither value met the criterion, the data were reported with appropriate qualifiers. Typical detection limits for pharmaceuticals in wastewater effluent were around 10 ng/L.

Table 2.2
Conditions used for analysis of beta-blockers

Compound	Retention time (min)	Parent ion	Product ion	Voltage (V)
Metoprolol	20.0	284	129	1.0
Propranolol	21.5	284	129	1.0
Hexachlorobenzene	12.8	142, 249, 284*		

*Hexacohlorbenzene quantified by SIM.

Validation of Methods for Acidic Drugs and Beta-Blockers

During initial phase of method development, the recovery of acidic drugs was often less than 50% when spike recovery experiments were performed in wastewater effluent or in deionized water. To identify the reason for the incomplete recoveries, a series of experiments were conducted. These experiments indicated that poor recoveries were due to the type of solid phase extraction resin used in the experiments and not due to losses due to adsorption on container walls or volatilization during blowdown. The findings of these experiments are summarized in the following paragraphs.

To assess possible losses during solvent transfer steps after derivatization, a set of samples containing 1 mg/L of the different acidic drugs was subjected to three different treatments after derivatization. In the first treatment, the extracts were blown to dryness with a gentle stream of nitrogen gas and resuspended in isooctane. In the second treatment, the samples were analyzed immediately after derivatization (no blowdown). In the third treatment, 250 μL of isooctane was added to the derivatized sample prior to nitrogen blowdown. These samples then were blown down until only the isooctane remained. Results indicated good recoveries for all samples and no significant differences between the three treatments (Figure 2.1). Similar experiments with volatile compounds, such as caffeine, showed losses of approximately 30% when samples were blown to dryness (data not shown). As a result of these experiments, we concluded that low recoveries of acidic drugs were not attributable to volatilization of the derivatives during blowdown.

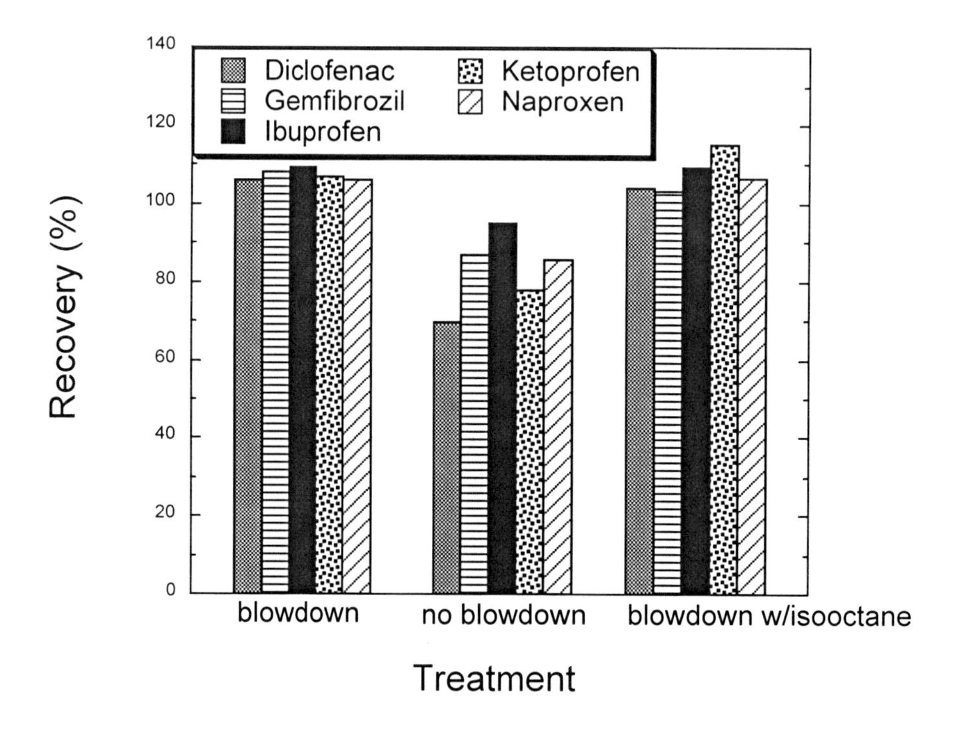

Figure 2.1 Effect of different blowdown methods on recovery of acidic drugs. Data are based on the average of duplicate experiments conducted with 1 mg/L of acidic drugs

To assess the possibility that acidic drugs were lost when the methanolic extracts were evaporated and transferred to 1-mL volumetric flasks (prior to derivatization), we added acidic drugs to 10 mL of methanol and treated the sample exactly as we would treat the eluent from SPE. Results indicate recoveries of approximately 80% for all of the analytes (Figure 2.2). Therefore, we concluded that significant losses of acidic drugs did not occur during blowdown and solvent transfer of the methanolic extracts.

After determining that significant losses of acidic drugs were not resulting from transfer steps, solvent evaporation or derivatization, we considered the possibility that solid phase extraction was responsible for the relatively low recoveries observed in many samples. We suspected that poor recoveries of the acidic compounds were attributable to the use of a mixed solid phase consisting of C-18 and Lichrolut ENV, which was the SPE used in the original method that we had used to develop our approach. To test this hypothesis, we extracted three sets of samples: (1) mixed resin SPE consisting of C-18 and Lichrolut EN; (2) single-resin SPE method consisting of only Lichrolut EN; and, (3) single-resin SPE method consisting of only C-18. Results suggested that the single resin C-18 SPE method yielded superior results compared to the other two methods (Figure 2.3). Therefore, we concluded that the low and variable recoveries of acidic drugs were attributable to the Lichrolut EN.

We also conducted an experiment to test the effect of different types of storage containers. Studies were conducted by adding 1,000 ng/L of each pharmaceutical to PFE-lined sample containers and to glass sample containers to which 2 g/L of sodium chloride was added prior to extraction. Results indicated no significant losses of acidic drugs in either type of

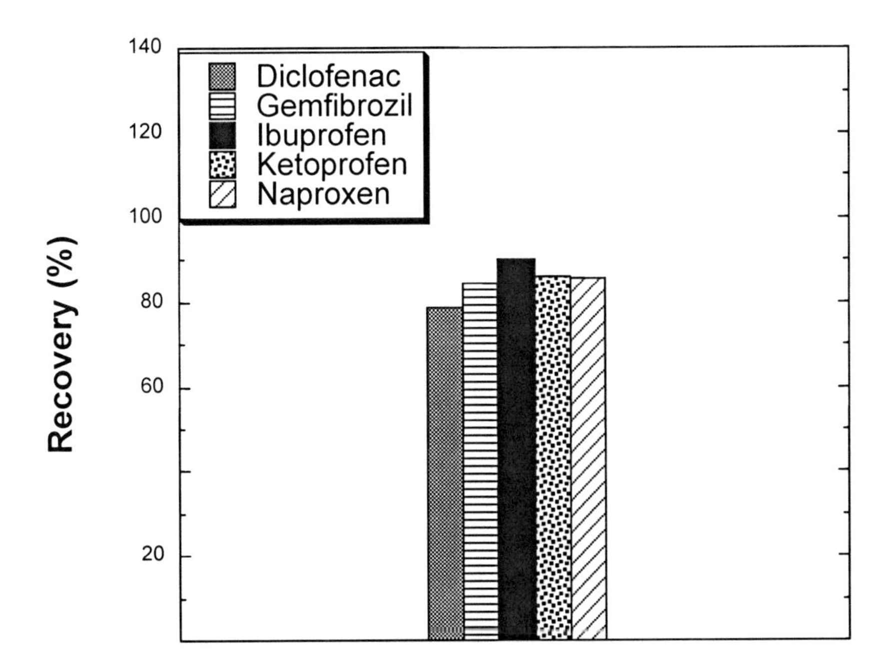

Figure 2.2 Recovery of acidic drugs from methanol stock solutions. The initial concentration of each compound was 100 μg/L. Results indicate average of duplicate extractions

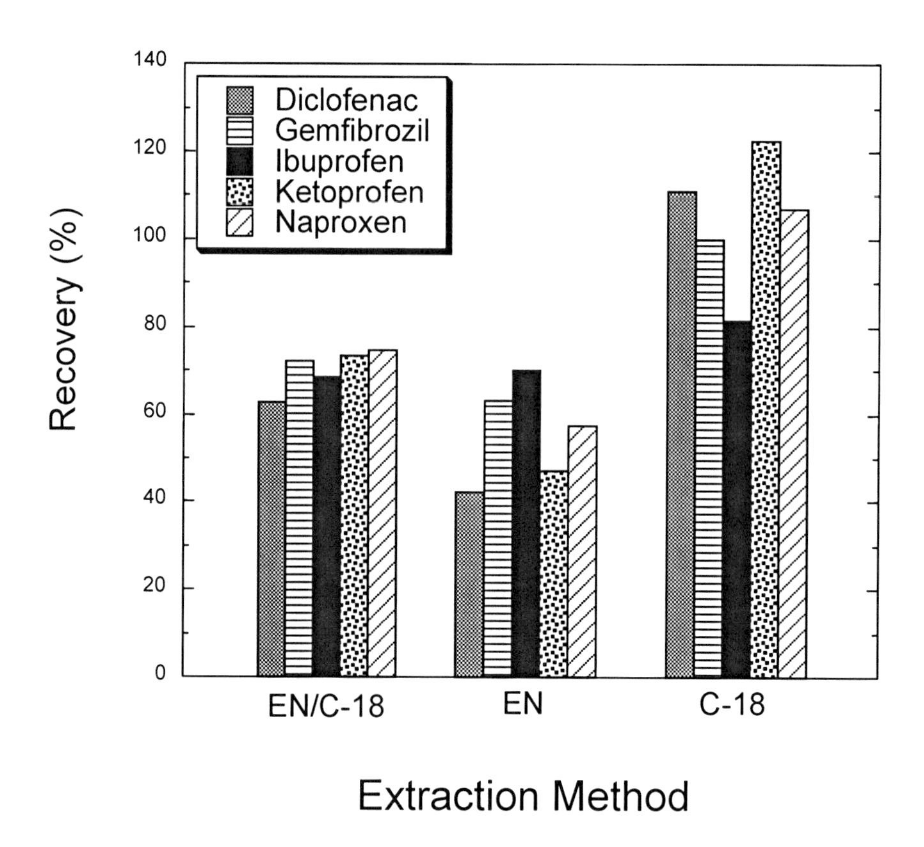

Figure 2.3 Recovery of acidic drugs subjected to extraction with different types of resins EN/C-18 = mixed resin; EN = Lichrolut EN resin; C-18 = C-18 based resin.

sample container (Figure 2.4). However, significant loss of metoprolol was observed. Unfortunately, the derivatization efficiency for metoprolol was low on this date, and the samples from the glass container only exhibited around 40% recovery. Repetition of this experiment on dates when recovery was better verified that significant losses of beta-blockers occurs through sorption (data not shown). The conditions used here likely overestimate the importance of sorption in environmental samples because other solutes in natural waters will compete for sorption sites. Nevertheless, we concluded that only glass containers should be used for sample collection and storage if beta-blockers are to be analyzed.

Following the completion of method development activities, we conducted a series of spike recovery studies using surface water collected from the West Central Basin Advanced Wastewater Treatment Plant in Los Angeles, California, the Sweetwater Groundwater Recharge Basin, near Tuscon, Arizona and the Russian River in Marin County, California. The samples from the West Central Basin Facility were analyzed for acidic compounds using the mixed C-18/Lichrolut-EN SPE while the other two sets of samples were analyzed using the C-18 SPE

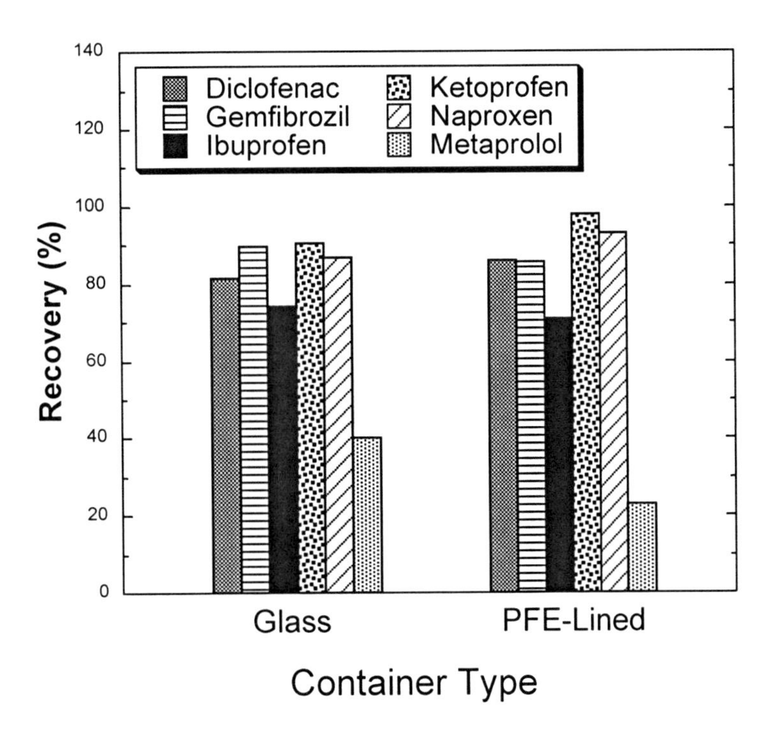

Figure 2.4 Effect of container type on recovery of acidic drugs and metoprolol. Results are based on the average of duplicate experiments.

only. Beta-blockers were extracted using the C-18 SPE method described previously. To assess recoveries, selected samples and deionized water blanks were amended with approximately 1,000 ng/L of each analyte. Recoveries of acidic compounds were consistent with our hypothesis that the mixed-resin SPE material was responsible for the variable recoveries depicted in Figure 2.3. After changing to the single-resin SPE, 85% of the measurements would have been in the acceptable recovery range. Average recoveries are plotted in Figures 2.5 through 2.7. The average difference between duplicate spike recovery or unspiked samples was approximately 15%, after excluding those samples where unusually low recoveries were measured. The low recovery of naproxen in the Caisson sample was due to the unexpected presence of chlorine inthe sample. Subsequent analysis of samples from the site showed acceptable recoveries when chlorine was quenched prior to analysis.

Comparsion of the results from the spike recovery experiments and the recovery of the internal standard indicate that mecoprop is an acceptable internal standard. The recovery of the mecoprop surrogate standard was reasonably well correlated (i.e., r^2=0.6) with the recovery of other acidic compounds in the samples (Figure 2.8) and no bias was observed. This relationship is comparable to the recoveries observed for other internal standards.

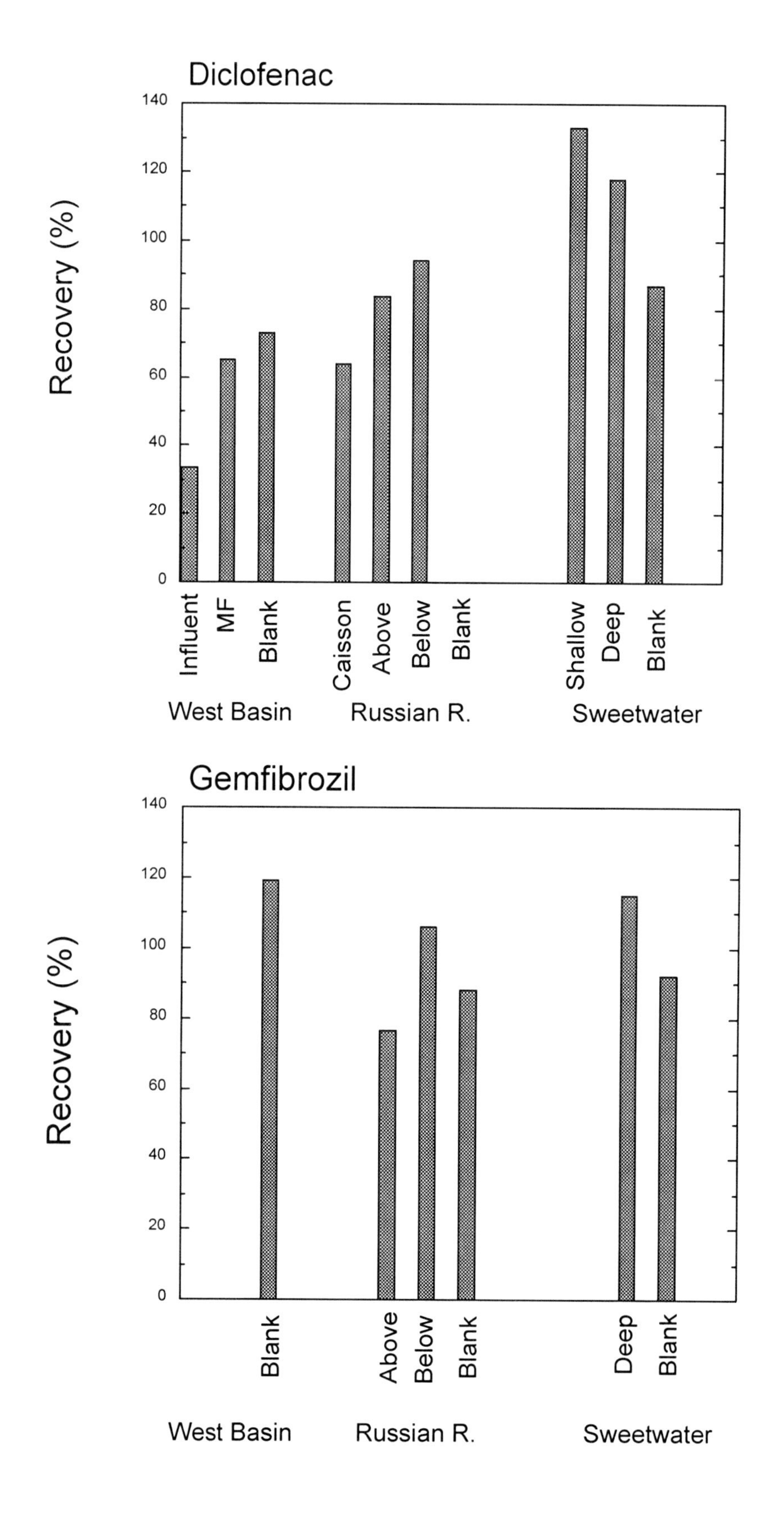

Figure 2.5 Recoveries of diclofenac and gemfibrozil in spike recovery samples

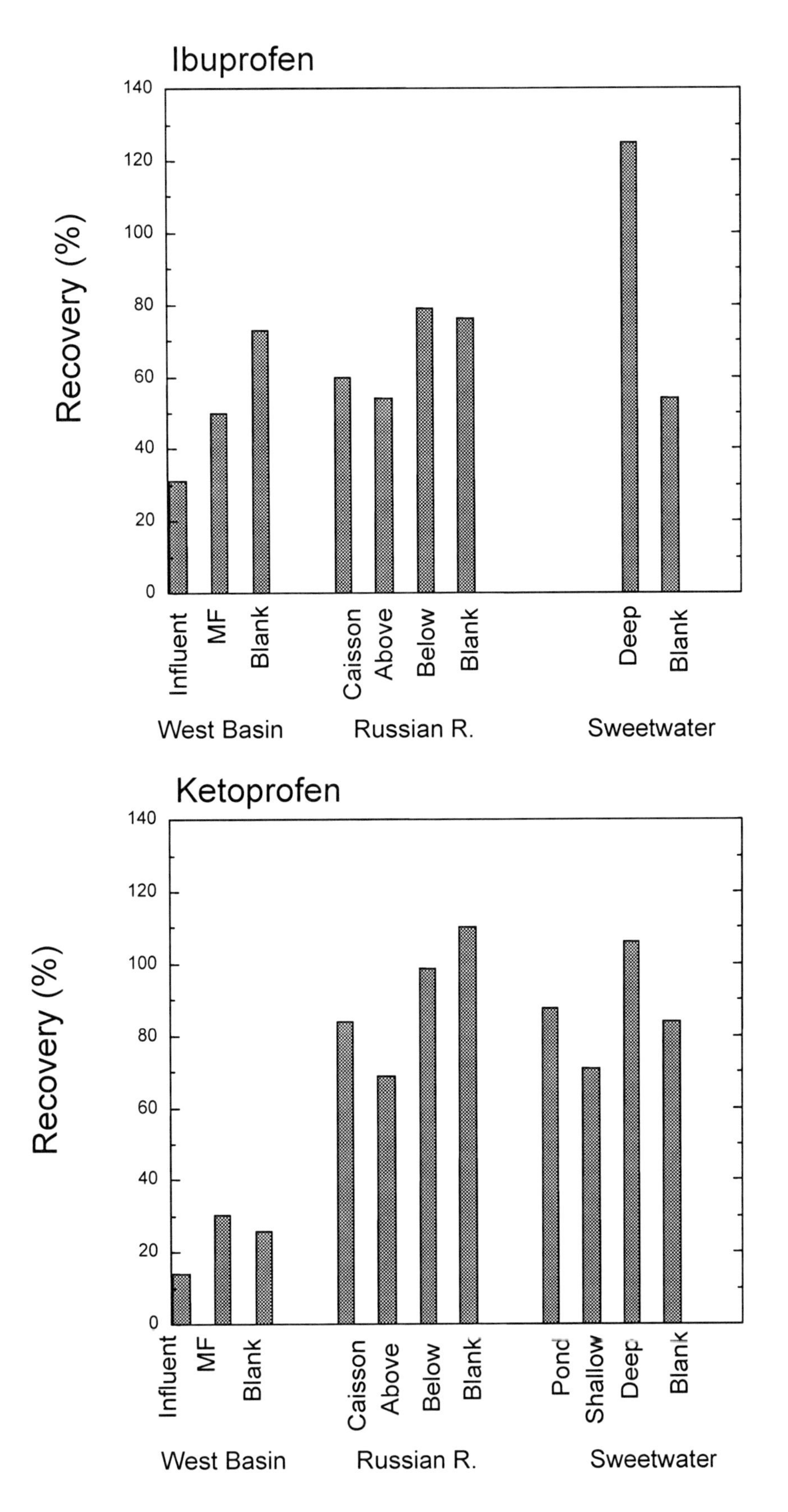

Figure 2.6 Recoveries of ibuprofen and ketoprofen in spike recovery samples

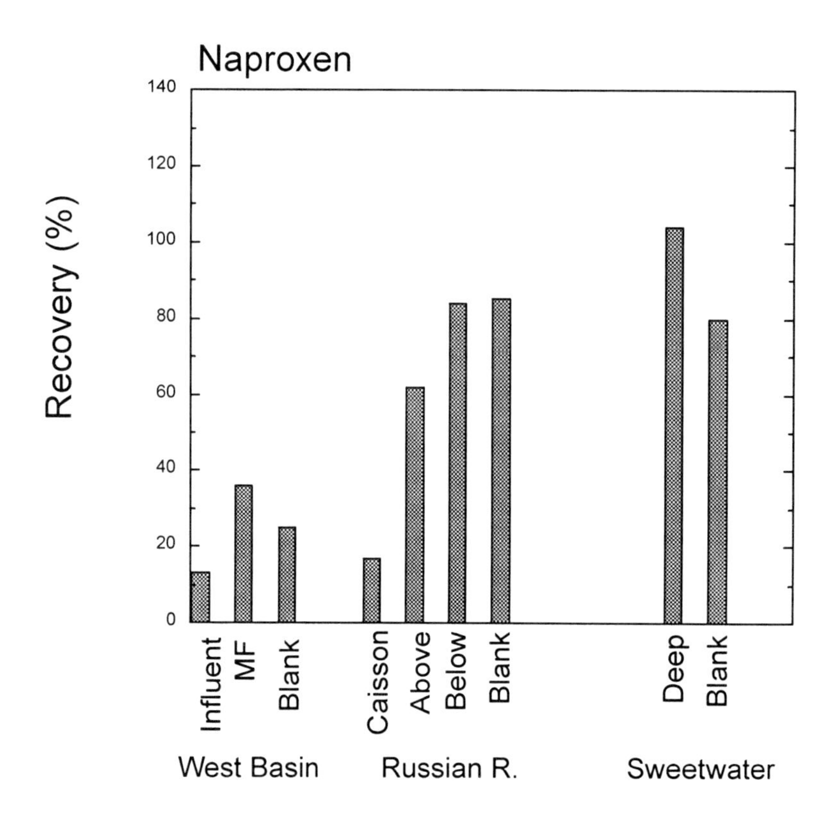

Figure 2.7 Recoveries of naproxen measured in spike recovery samples

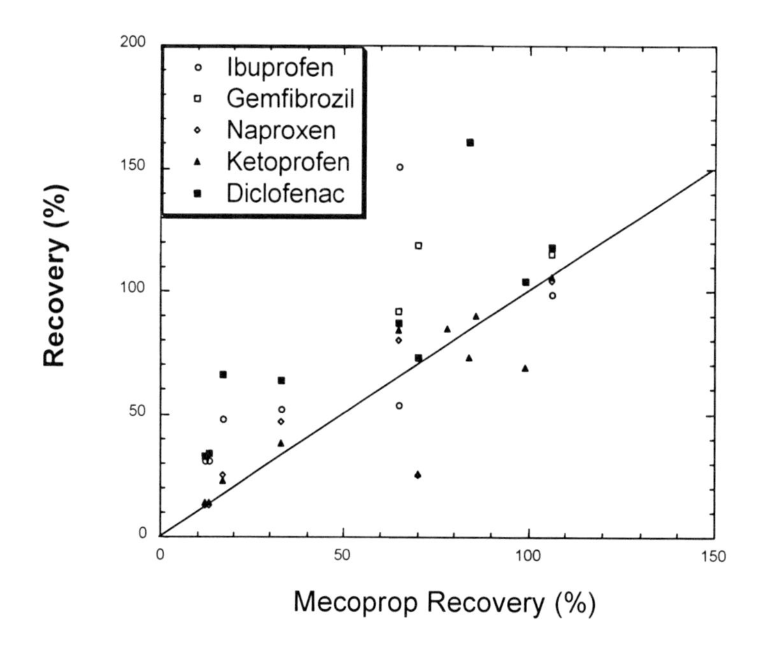

Figure 2.8 Relationship between the recovery of mecoprop and acidic drugs. The line indicates a 1:1 relationship.

Although the internal standard method provided a convenient way of assessing recoveries of the acidic drugs, no comparable approach could be perfected for the beta-blockers, despite numerous attempts to identify suitable internal standards. Attempts to use labeled propranolol as an internal standard failed because the compound was labeled on the aromatic ring and that portion of the molecule could not be used for MS/MS analysis. A supplier of a form of labeled propranolol with the label on the side chain was identified eventually. However, the manufacturer was unable to provide the compound because they did not have it in stock. Subsequent inquiries indicated that the company would sell us the compound if we were willing to purchase a relatively large quantity.

We also explored alternative means of including an internal standard in the beta-blocker method. Initially, we attempted to use selected-ion monitoring (SIM) to discriminate between labeled and unlabeled forms of propranolol. However, the method would have reduced our sensitivity for propranolol and prevented analysis of the compound in wastewater effluent samples. After we determined that labeled propranolol would not be useful as an internal standard, we investigated commercially available compounds with structures similar to that of the beta-blockers. We determined that epinephrine and deoxyepinephrine might be acceptable internal standards because they have similar structures to the beta-blockers but are not excreted by humans in significant quantities. To test these compounds, we developed a GC/MS/MS methods for the derivatives of both compounds. The sensitivity and linear range for both compounds were similar to that observed for the beta-blockers. However, attempts to extract the compounds from water failed, presumably because the compounds are more polar than the beta-blockers.

As a result of our inability to identify a suitable internal standard for the beta-blockers, we used the results of spike recovery samples to monitor the efficiency of the analytical procedure. Results from spike recovery studies conducted with the beta-blockers are depicted in figures 2.9 and 2.10. These results indicated that recoveries of the beta-blockers were usually below the 60% target. Continued practice by the analyst responsible for the measurements eventually resulted in recoveries above 60%, as shown in the next chapter. However, these results emphasize the importance of using great care when measuring beta-blockers.

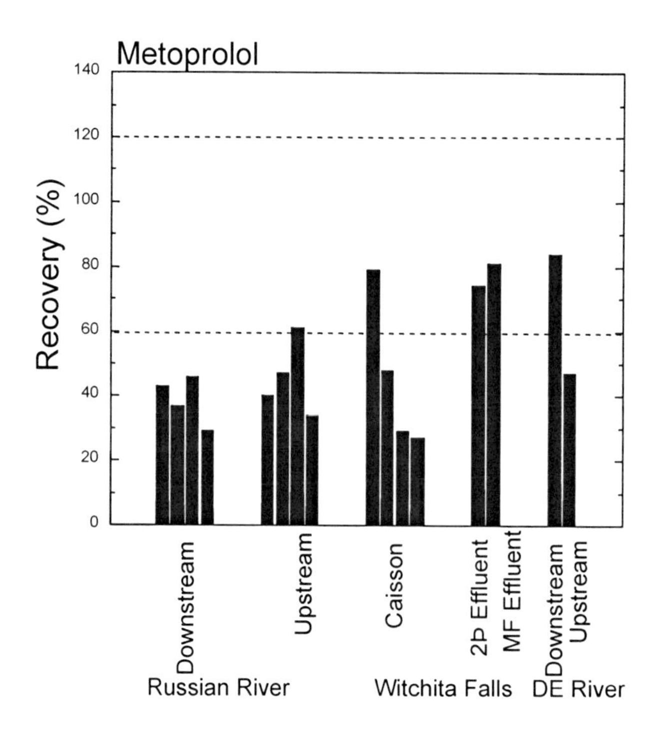

Figure 2.9 Recoveries of metoprolol

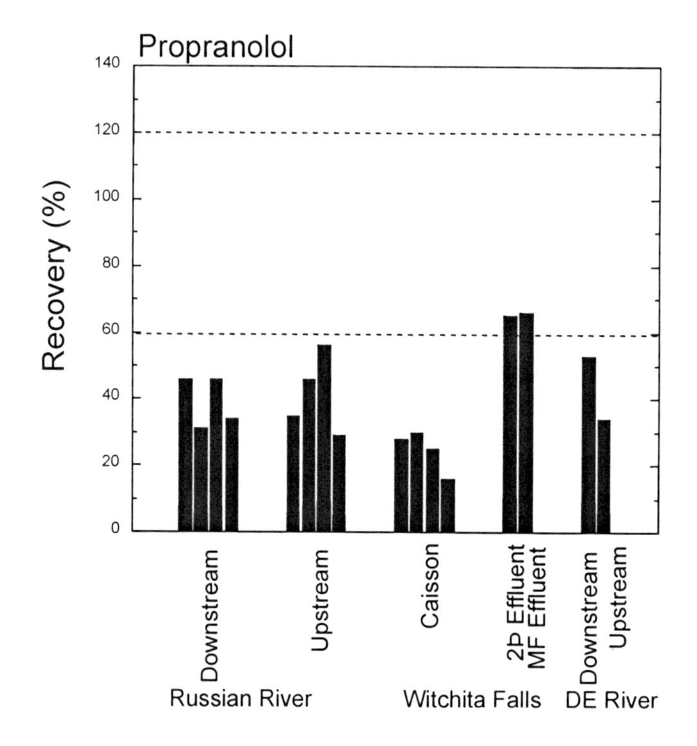

Figure 2.10 Recoveries of propranolol

ANTIBIOTICS

The analysis of antibiotics was performed by high performance liquid chromatography/mass spectrometry after solid phase extraction. As a result of matrix interference related to the organic matter that was co-extracted with the antibiotics, the instrument response varied between samples. Antibiotics were quantified using internal standards and standard additions to account of matrix effects. The following sections describe the final analytical methods adopted for the occurrence survey and the method development and validation studies performed in conjunction with these analyses.

Grab samples were collected in 1-L amber glass bottles with Teflon-lined screw caps. Each bottle was kept in an individual polyethylene bag. Prior to sampling, bottles were cleaned with Micro brand laboratory detergent, rinsed with water followed by methanol and deionized water between each analysis. Bottles were shipped to participating utilities in coolers with blue ice packs and were returned by overnight mail within one day of sample collection. The samples were then filtered through 0.5-µm glass fiber filters (Pall, Ann Arbor, MI). 0.1 M of NaCl was added. The addition of NaCl greatly improved the recoveries for the sulfonamides and trimethoprim by enhancing the salting out of these antibiotics. Addition of NaCl also helped stabilize the recoveries for the fluoroquinolones. The samples were then acidified to pH 2.5 with concentrated phosphoric acid.

For samples collected from wastewater treatment plants or water treatment plants using chlorine for disinfection, $Na_2S_2O_3$ (final concentration 2 mg/L) was added to the samples bottle as a preservative. Samples from other locations were collected in bottles without any added preservative. Each set of samples was shipped with a field blank, consisting of deionized water, which was analyzed with the samples. Samples were collected by field personnel who were familiar with trace organic sampling protocols. Field personnel wore polyethylene gloves when handling bottles and were instructed to minimize the amount of time that the bottle is kept uncapped outside of the cooler. Upon arrival at Georgia Tech, samples and log sheets will be visually inspected and transferred to a 5°C storage area. Samples were extracted as soon as practical and within no more than 72 hours after arrival.

Each set of 8-12 samples was extracted in a batch along with appropriate QA/QC standards. The following samples were included with each set of samples:

(1) Field blank (1 L of deionized water that traveled to and from the field site);
(2) Matrix recovery sample (1 sample from the site spiked with all analytes at 1,000 ng/L);
(3) Duplicate sample;
(4) Auxiliary standard consisting of a mixture of analytes.

Each 1-L sample was extracted through a 500-mg anion exchanger (Isolute, Mid Glamorgan, U.K.) stacked on top of a 500-mg hydrophilic-lipophilic balance HLB cartridge (Waters, Taunton, MA). Both cartridges were pre-conditioned with 6 mL of methanol followed by 6 mL of 4.38 mM H_3PO_4. The 1-L samples were extracted at a rate of approximately 6mL/min. The HLB cartridges were eluted with 10 mL of 95% methanol/5% 4.38 mM H_3PO_4. The analytes were eluted into high-density polyethylene conical test tubes (Becton Dickinson, Franklin Lakes, NJ). The analytes were then blown down in a water bath at 30° C using nitrogen. The analytes were reconstituted in 1 mL of 20% methanol /80% 4.38 mM H_3PO_4 containing 1 g/L of both internal standards (sulfamerazine and lomofloxacin). When standard addition was employed for quantification, a 300 μL aliquot of the extract was amended with a known amount of a mixed stock solution of the analytes. All samples were then transferred into amber vials for LC/MS analysis.

Analytes were injected into an Agilent 1100 series HPLC (Palo Alto, CA). A 2.1x150 mm 5-micron Zorbax SB-C18 column (Agilent, Palo Alto, CA) was used to separate the analytes. The column temperature was set at 30°C. Two mobile phases were used. Mobile phase A contained 1 mM ammonium acetate, 0.007% glacial acetic acid and 10% acetonitrile. Mobile phase B was 100% acetonitrile. A flowrate of 0.25 mL/min was used. The mobile phase gradient for this method is shown in Table 2.3. After the gradient was complete, the column was flushed with 100% B for 10 minutes. A post-time of 15 minutes was used to allow the column to equilibrate before the next sample injection.

Ions of the analytes were detected using a HP1100 Series mass selective detector (MSD) (Agilent, Palo Alto, CA). Positive mode electrospray ionization and selected ion monitoring were used. In order to avoid excessive cleaning of the capillary, the MSD was only operated in the period of 6 minutes to 27 minutes after sample injection. Analytes were detected at a fragmentor voltage of 85 V and 120 V. The run at the higher fragmentor voltage provided additional confirmation of the presence of the antibiotics. The retention times, molecular ions and confirming ions for the antibiotics are shown in Table 2.4. The relative abundance of the ions at the two fragmentor voltages used are shown in Table 2.5. The antibiotics were quantified using the standard addition method or the internal standard method. The molecular ion of each antibiotic was used in both quantification techniques.

Table 2.3
Mobile phase gradient for LC/MS analytical method

Time (min)	B (%)
0	0
2	0
8	8.5
20	18
25	50
30	100

Table 2.4
Retention time, molecular ion and fragment ions of antibiotics

Antibiotic	Retention Time (min)	$[MH]^+$ ion	Confirming ion 2	Confirming ion 3
Ciprofloxacin	17.2	332	314	288
Enrofloxacin	20.6	360	342	316
Lomefloxacin	17.9	352	334	
Norfloxacin	16.3	320	302	276
Ofloxacin	16.3	362	318	261
Sulfamerazine	9.5	265	156	
Sulfamethazine	12.6	279	156	
Sulfamethoxazole	18.4	254	156	
Trimethoprim	14.1	291	261	

Table 2.5
Relative abundance of molecular and confirming ions for the two fragmentor voltages

	[MH]+ ion Relative Abundance (%)		Confirming ion 2 Relative Abundance (%)		Confirming ion 3 Relative Abundance (%)	
	Fragmentor = 85 V	Fragmentor =120 V	Fragmentor = 85 V	Fragmentor =120 V	Fragmentor = 85 V	Fragmentor =120 V
Ciprofloxacin	100	100	4	71	9	73
Enrofloxacin	100	100	14	36	9	90
Lomefloxacin	100	100	2	25		
Norfloxacin	100	97	5	100	13	90
Ofloxacin	100	92	10	100	1	33
Sulfamerazine	100	52	15	100		
Sulfamethazine	100	100	5	71		
Sulfamethoxazole	100	37	41	100		
Trimethoprim	100	100	1	9		

MS conditions: drying gas flowrate=10 mL/min; drying gas temperature=350°C; nebulizer pressure 30 psig; capillary voltage=3500 V.

Validation of Methods for Antibiotics

After developing analytical methods for the antibiotics, a series of experiments were performed to test and improve the method performance. These method validation studies are described in the following paragraphs.

Recovery studies were performed to identify steps in which analyte loss could occur. To determine the potential for loss of analytes during sample blowdown, 10 mL of methanol was spiked with 250 ng/L to 100 μg/L of antibiotics and blown to dryness prior to redissolving the compounds in 1 mL of a methanol/water mixture. The experiments were performed in various high-density polyethylene and glass containers. Analyte losses during blowdown were less significant for sulfonamides than for fluoroquinolones (Figure 2.11). Among the tested containers, high-density polyethylene conical tubes yielded the best and most consistent recoveries for ciprofloxacin and were used for all subsequent experiments in SPE eluent collection and blowdown.

Experiments also were conducted to assess the stability of antibiotics during storage. Results of these analyses indicated that fluoroquinolones remain stable for a longer period in acidic solutions. Therefore, to minimize compound losses prior to solid phase extraction, samples were acidified with phosphoric acid immediately after the samples were filtered with 0.5 μm glass-fiber filters. In addition, the analytes were redissolved after blowdown in an acidified water and methanol mixture (800 μL/200 μL) and were transferred to amber vials to minimize breakdown of antibiotics prior to analysis by LC/MS. It should be noted that several of the antibiotics are susceptible to photolysis, especially upon exposure to sunlight. Therefore,

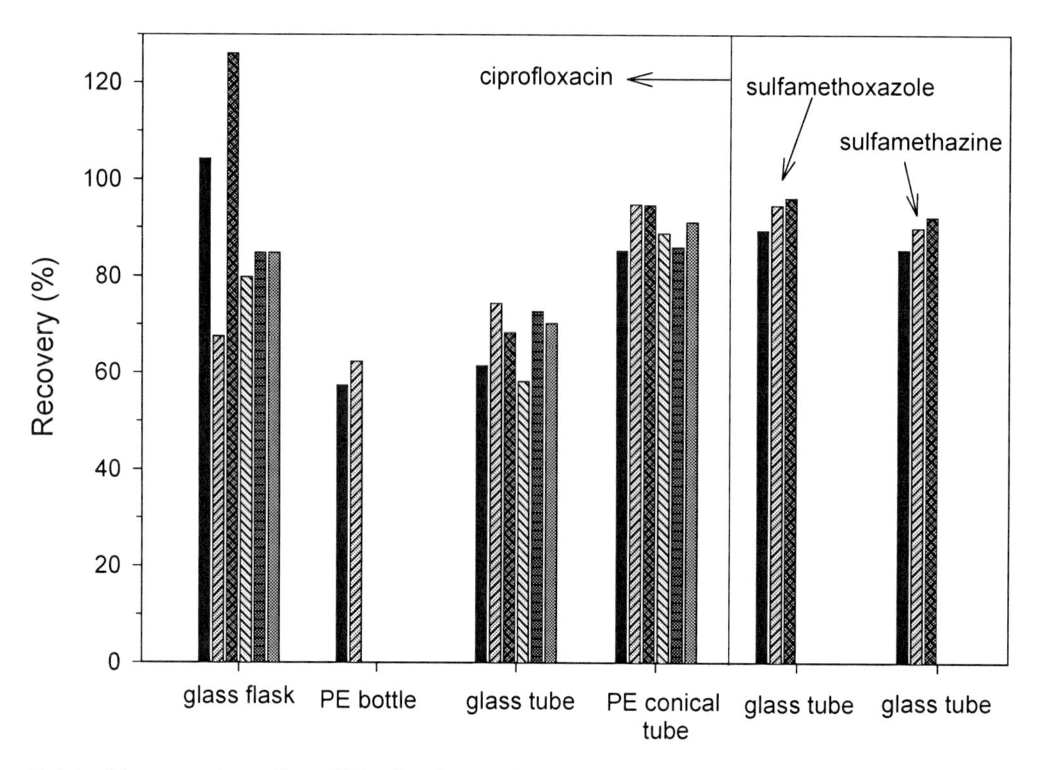

Figure 2.11 Recoveries of antibiotics in various containers during blowdown. Tested containers: borosilicate volumetric flasks (25 mL), polyethylene bottles (100 mL), borosilicate test tubes, and high-density polyethylene conical tubes.

samples and extracts should be stored in amber bottles or in clear glass bottles wrapped with aluminum foil. In cases in which samples could not be immediately analyzed by LC/MS, the vials were stored at 0°C until analysis.

Studies also were performed to assess the effect of the wastewater matrix on sample recoveries. In these studies, the recoveries were calculated by both the internal standard method and the standard addition method. Use of standard addition method is appropriate since all the antibiotics exhibit linear calibration curves within the investigated concentration range ($r^2 > 0.98$). Although the standard addition method is more time consuming than the internal standard method, it tends to yield more accurate results if only one internal standard is used for a variety of compounds.

To assess the need for using the standard addition method, a series of recovery experiments were conducted by adding 1,000 ng/L of each antibiotic to deionized water samples and to in samples from conventional and advanced wastewater treatment plants. For the fluoroquinolones, ciprofloxacin, norfloxacin, enrofloxacin, and ofloxacin, the average recoveries for the standard addition method (±standard deviation) were 105±55%, 103±61%, 86±42% and 113±71%, respectively (Figures 2.12-2.15). For the internal standard method, the recoveries were 101±31%, 97±28%, 92±25% and 116±23%, respectively. Overall, the method is robust and reliable for the analysis of fluoroquinolones in different wastewater matrices. The internal standard and standard additions methods yield comparable results; however, standard additions yielded higher variability in recoveries, presumably because of human error.

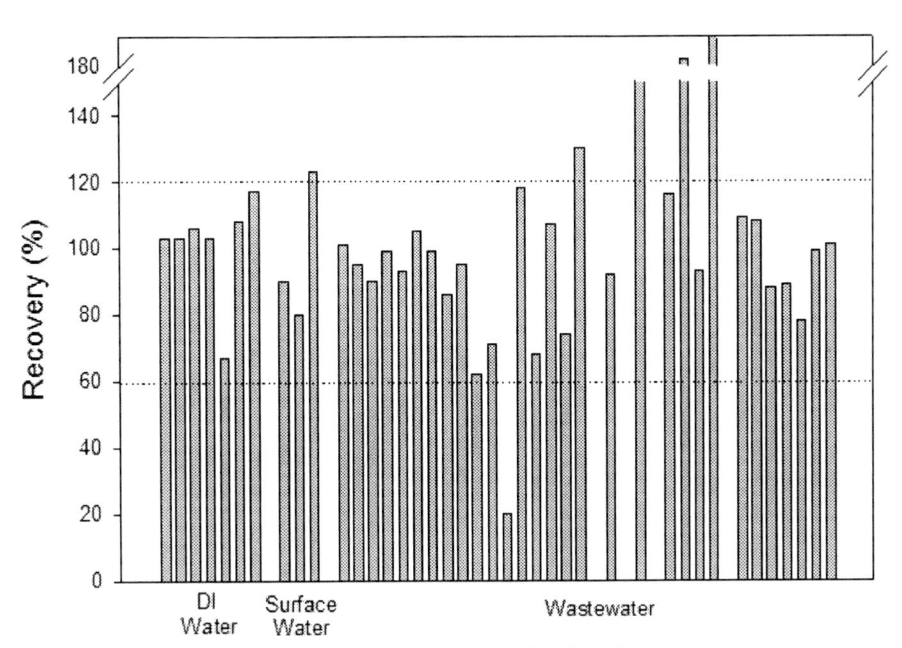

Figure 2.12 Recoveries of ciprofloxacin in deionized water and wastewater effluent. All recoveries were calculated based on the internal standard quantification method in which lomefloxacin was the internal standard.

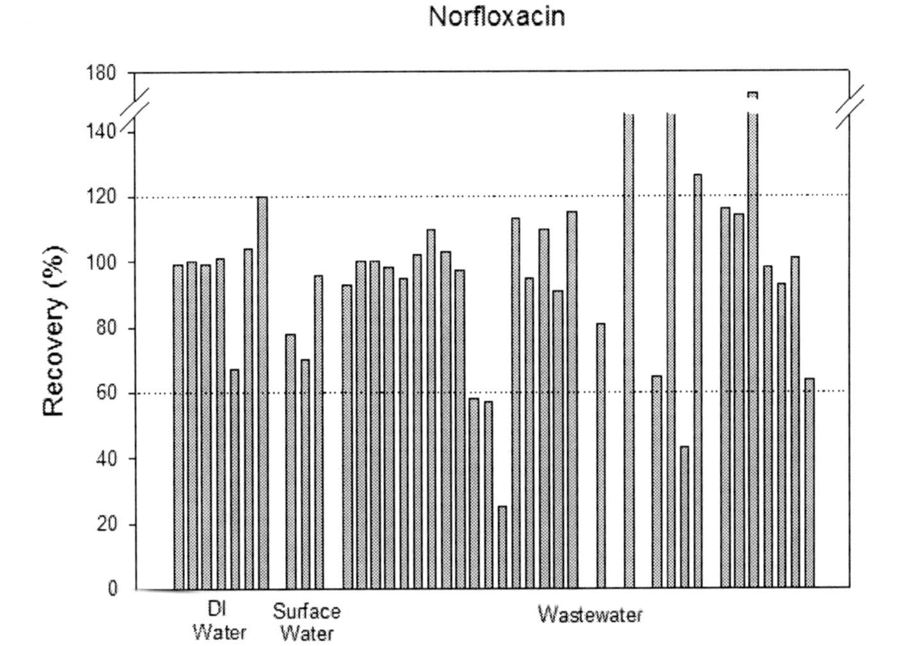

Figure 2.13 Recoveries of norfloxacin in deionized water and wastewater effluent. All recoveries were calculated based on the internal standard quantification method in which lomefloxacin was the internal standard.

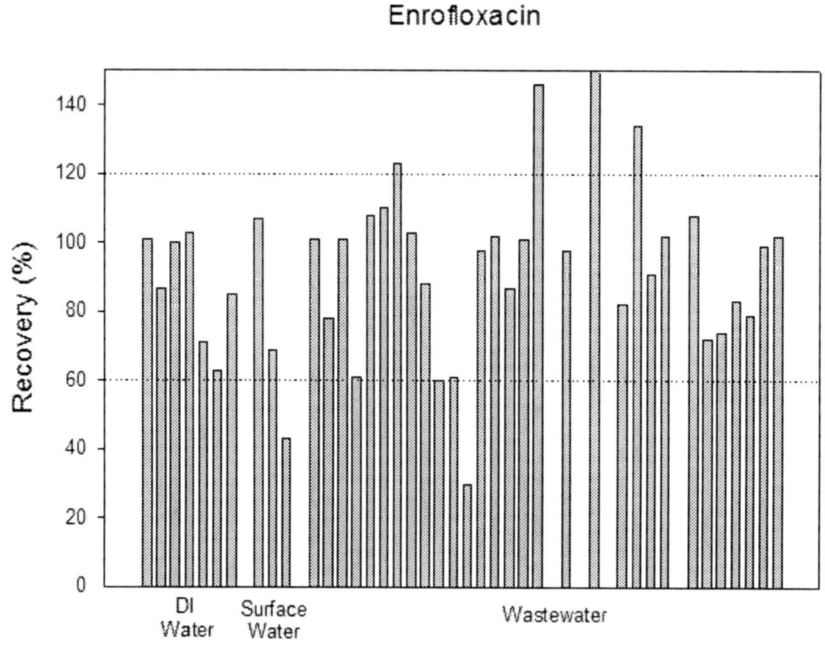

Enrofloxacin

Recovery (%)

DI Water
Surface Water
Wastewater

Figure 2.14 Recoveries of enrofloxacin in deionized water and wastewater effluent. All recoveries were calculated based on the internal addition method in which lomefloxacin was the internal standard.

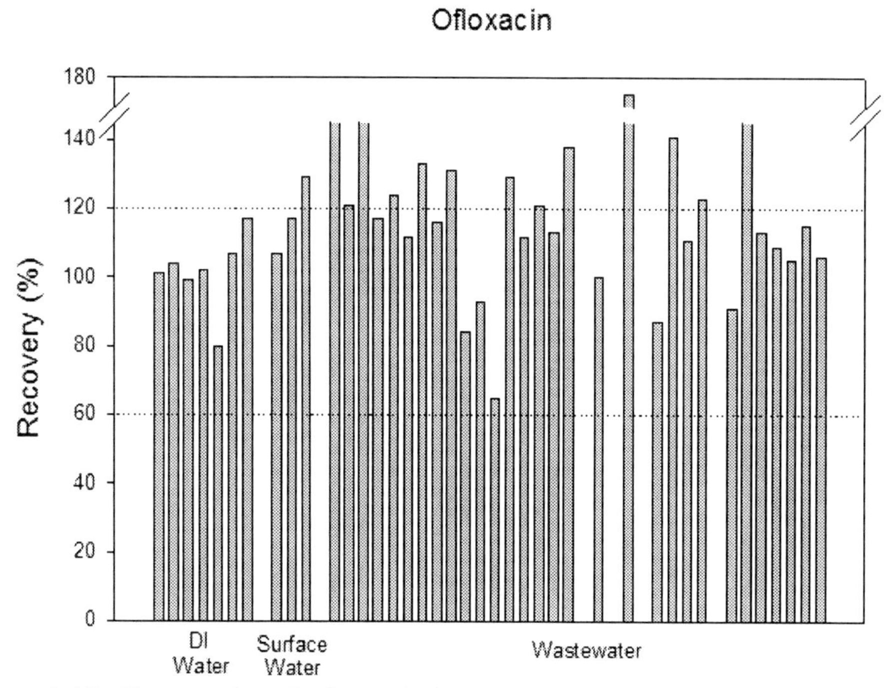

Figure 2.15 Recoveries of ofloxacin in deionized water and wastewater effluent. All recoveries were calculated based on the internal standard quantification method in which lomefloxacin was the internal standard.

The recoveries for the sulfonamides were generally lower than those of fluoroquinolones. The average recoveries of sulfamethoxazole and sulfamethazine were 46±34% and 33±29%, respectively by standard additions method, and were 51±31% and 50±38%, respectively by internal standard method (Figures 2.16 and 2.17). The average recovery of trimethoprim was 100±55% for the standard additions method, which was significantly different from the value of 191±141% obtained by the internal standard method (Figure 2.18).

The discrepancies between the standard addition method and the internal standard method were especially evident for trimethoprim. The relatively high recoveries for trimethoprim most likely are attributable to two factors. First, the use of sulfamerazine as an internal standard when quantifying by the internal standard method introduces error because this internal standard is more similar to the sulfonamides than it is to trimethoprim. As a result, the internal standard quantification overestimates the concentrations of trimethoprim. Secondly, the co-elution of organic matter in the sample results in considerable amount of signal suppression. Figures 2.19 and 2.20 illustrate the signal suppression effect for the antibiotics in samples from the Sweetwater Recharge Facility. The signal suppression is defined as the percent decrease in signal intensity for an antibiotic in a sample extract versus in a deionized water matrix and was calculated using equation 2. In those samples, the internal standards were assumed to be absent in the unspiked samples.

$$\text{Signal Suppression (\%)} = \left(1 - \left[\frac{I_S - I_X}{I_{DI}}\right]\right) * 100 \qquad \text{(Eq. 2)}$$

where I_s is the signal intensity of antibiotic in the spiked solution of the matrix of interest, I_X is the signal intensity of antibiotic in unspiked solution of the matrix of interest, I_{DI} is the signal intensity in deionized water spiked with the same amount of antibiotic as the matrix of interest. The volume changes were also taken into account.

The UV_{254} absorbance (a surrogate measure of the concentration of dissolved organic carbon DOC) for the deep well, shallow well and pond samples prior to extraction were 0.028, 0.267 and 1.117, respectively. Clearly, signal suppression increases with increasing DOC in the sample and the difference in signal suppression between the analytes and the internal standards exists in all matrices. Figure 2.19 indicates that trimethoprim typically exhibits a lower susceptibility toward signal suppression than the sulfonamides. As a result, the concentration of trimethoprim is overestimated based on the sulfamerazine internal standard quantification. If the approximately 35% overestimation due to signal suppression is taken into account for trimethoprim, the average recovery of trimethoprim (159%) would be closer to 103% (i.e., 159%×65%).

The difference in signal suppression between lomefloxacin and the other fluoroquinolones is smaller (Figure 2.20). This is likely due to the fact that the structures and retention times of fluoroquinolones are more similar to each other. However, ofloxacin and norfloxacin exhibit lower signal suppression than lomefloxacin and thus their recoveries may also be overestimated.

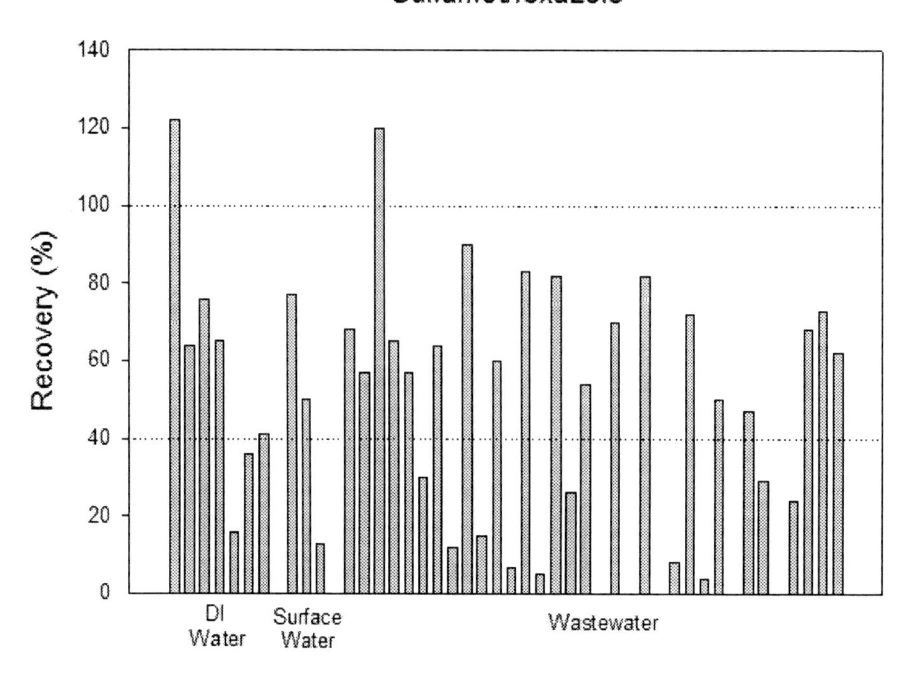

Figure 2.16 Recoveries of sulfamethoxazole in deionized water and wastewater effluent. All recoveries were calculated based on the internal standard quantification method in which sulfamerazine was the internal standard.

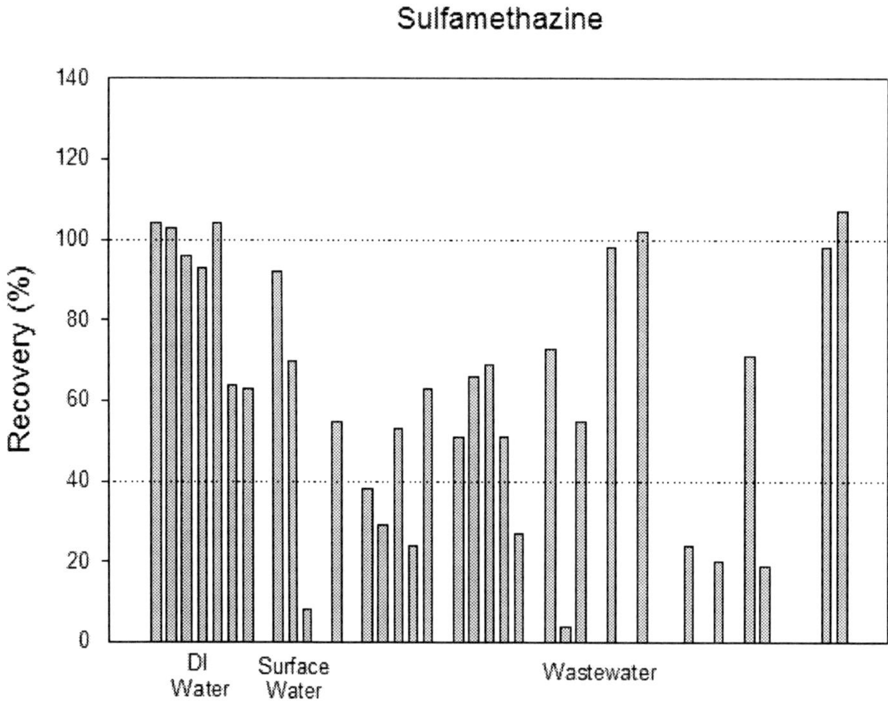

Figure 2.17 Recoveries of sulfamethazine in deionized water and wastewater effluent. All recoveries were calculated based on the internal standard quantification method in which sulfamerazine was the internal standard.

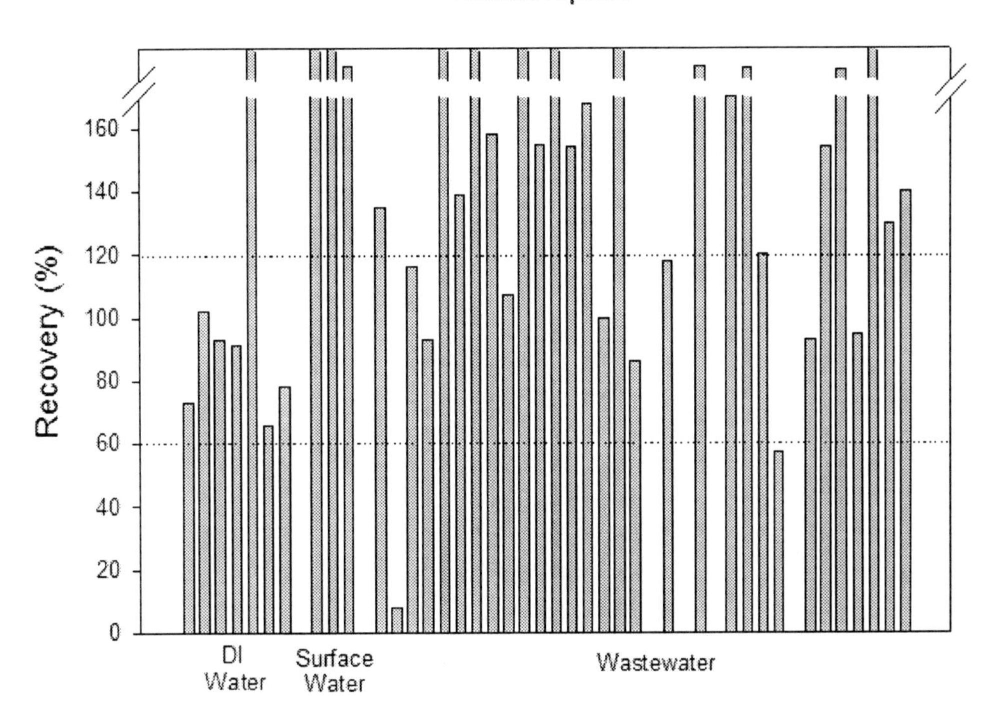

Figure 2.18 Recoveries of trimethoprim in deionized water and wastewater effluent. All recoveries were calculated based on the internal standard quantification method in which sulfamerazine was the internal standard.

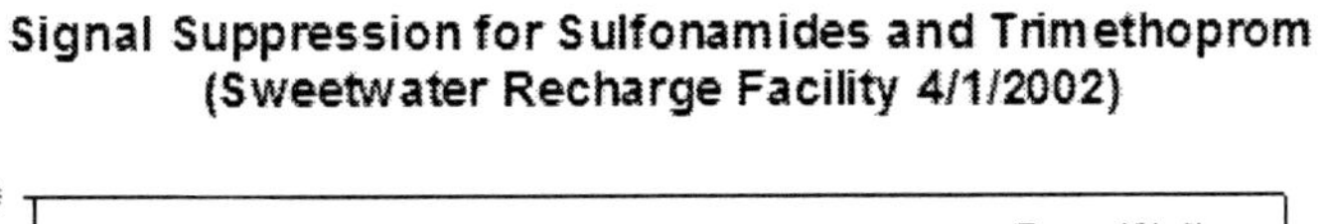

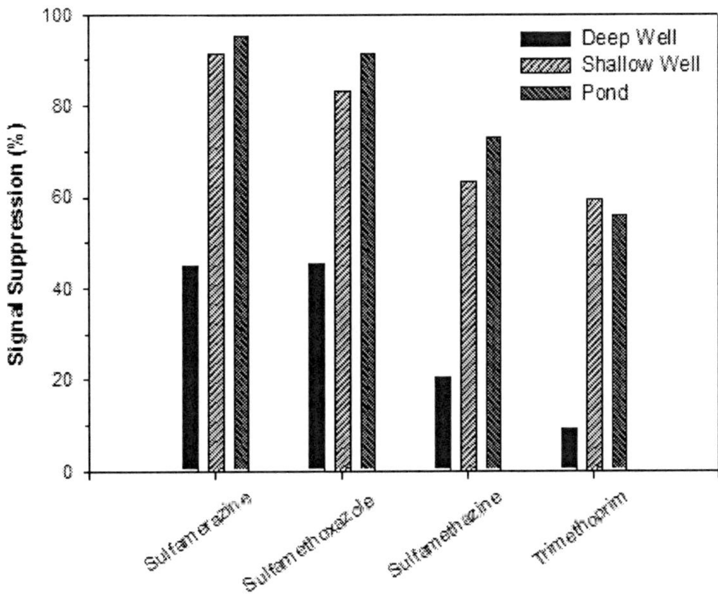

Figure 2.19 Signal suppression for sulfonamides and trimethpoprim caused by the matrix effect from the Sweetwater Recharge Facility

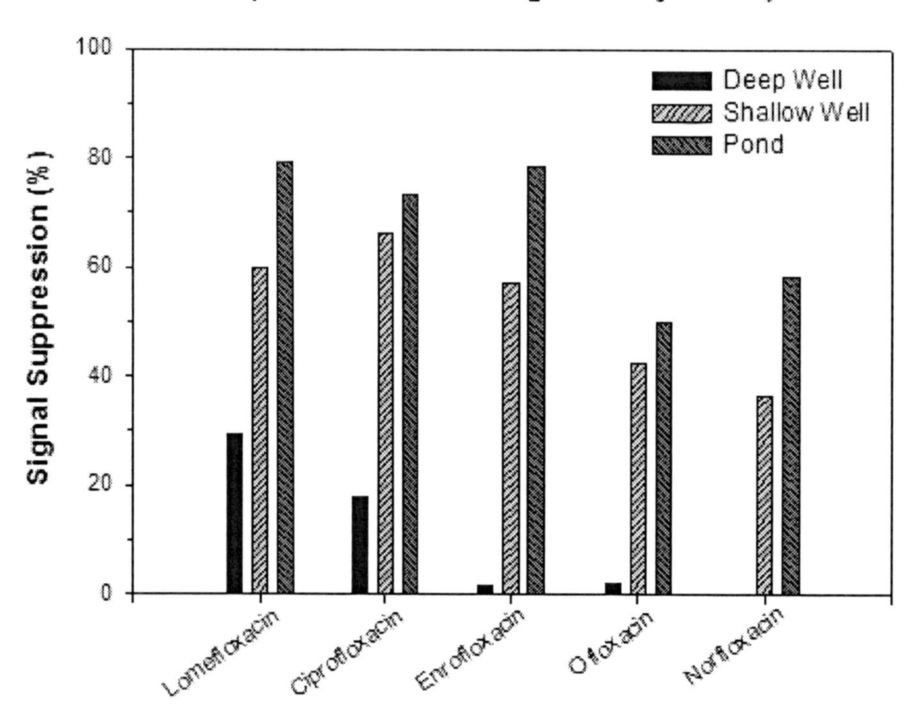

Figure 2.20 Signal suppression for fluoroquinolones caused by the matrix effect from the Sweetwater Recharge Facility.

SUMMARY OF METHOD PERFORMANCE

The analytical methods described in this chapter are capable of detecting PhACs in wastewater effluent, surface waters and groundwater. Although the use of these methods requires careful attention to sample handling and QA/QC procedures, analysts with training in modern analytical techniques can achieve sensitive detection limits and reproducible results. The typical detection limits for the acidic drugs and beta-blockers are approximately 10 ng/L while typical detection limits for the antibiotics range from 30 to 50 ng/L, depending on the matrix. The acceptable recovery range for all of the PhACs is 60-120%, based upon the recovery of the surrogate standard, mecoprop, for the acidic drugs and spike recoveries for the beta-blockers and antibiotics. As illustrated in the following chapter, recoveries in this range could not always be achieved, and in some cases, data outside of this range are reported with appropriate qualifiers for informational purposes.

CHAPTER 3
OCCURRENCE SURVEY

SITE DESCRIPTIONS

As mentioned previously, the main sources of PhACs in the aquatic environment are municipal wastewater and waste from agricultural operations involving large numbers of animals. PhACs originating in human therapy are most likely to be present in waters that receive large discharges of wastewater effluent or are subject to sewage leaks or combined sewer overflows. With respect to water supplies where intentional or unintentional reuse of municipal wastewater effluent occurs, the potential occurrence of PhACs may be especially important. Furthermore, the occurrence of PhACs in municipal wastewater effluent is easier to study than PhACs related to agricultural runoff or combined sewer overflows because runoff and overflow events vary considerably with hydrologic conditions. As a result, we focused our efforts on sampling of municipal wastewater effluent and water recycling systems that serve as important barriers to the entry of wastewater-derived contaminants into drinking water sources.

As part of the site selection process, we identified sites where it would be possible to obtain representative samples from both conventional and advanced wastewater treatment plants. Sites also were identified where samples could be collected after the discharge of effluent into aquifers or engineered treatment wetlands. Several surface waters also were included to serve as reference sites for background samples. A final list of sites sampled during the occurrence survey is included in Table 3.1. The selected sites included a total of eight conventional wastewater treatment plants, three advanced wastewater treatment plants, two engineered treatment wetlands and three background sites. Each of the sites is described briefly below.

R.M. Clayton Water Reclamation Center (Atlanta, GA)

The R.M. Clayton water reclamation center is a 5.27 $m^3 s^{-1}$ (122 MGD) facility. The plant is equipped with primary treatment, followed by activated sludge treatment with three-stage biological phosphorous removal in activated sludge reactors. Following clarification, the wastewater undergoes granular media filtration and ultraviolet (UV) disinfection.

Dublin/San Ramon Advanced Wastewater Treatment Plant (Dublin, CA)

The Dublin/San Ramon Services municipal wastewater treatment plant is a 0.50 $m^3 s^{-1}$ (12 MGD) facility. The plant is equipped with primary screening and clarification, followed by activated sludge treatment and chlorine disinfection.

Hyperion Wastewater Treatment Plant (Los Angeles, CA)

The Hyperion municipal wastewater treatment plant treats a total of 15.7 $m^3 s^{-1}$ (358 MGD) of municipal wastewater effluent with advanced primary treatment or secondary treatment. The water sampled during the occurrence study originated in the secondary treatment plant, which treats 8.45 $m^3 s^{-1}$ (193 MGD) of wastewater effluent. The secondary treatment plant

Table 3.1
Summary of sample collection sites in the occurrence survey

Location*	Description	Dates Sampled
Conventional WWTPs		
Clayton	BNR, UV	3/19/02, 8/20/02
Dublin/San Ramon	AS, Cl_2	3/21/02
Hyperion	AS, Cl_2	9/18/01, 5/22/02
Mt. View	AS, biotower, UV	9/4/01, 4/9/02, 8/21/02
Roger Road	AS, Cl_2	2/17/02, 4/1/02
San Jose/Santa Clara	BNR, effluent filtration, Cl_2	3/26/02, 6/26/02
South Cobb	AS, Cl_2	2/15/02, 6/12/02
Treatment Plant S	O_2-AS, Cl_2	7/1/02, 10/5/02
Advanced Treatment Plants		
F. Wayne Hill	Activated carbon, ozone	4/22/02,7/17/02, 10/17/02
Treatment Plant T	Microfiltration, RO, UV	9/18/01
West Basin	Microfiltration, RO, UV	9/18/01, 5/22/02, 9/10/02
Groundwater Recharge		
Sweetwater Recharge Facility	Secondary effluent recharge	2/17/02, 4/1/02
Engineered Wetlands		
Mt. View	Tertiary effluent, HRT ~7 days	9/4/01,4/9/02, 10/15/02
Prado	Effluent-dominated river	4/6/02
Background		
MWD Water	Los Angeles water supply	9/18/01
Russian River	Marin County, CA	5/14/01, 6/11/01, 8/13/01
Chatahoochie River	Intake for Quarles WTP, GA	12/12/02, 12/19/02
Lake Altoona	Intake for Wyckoff WTP, GA	9/6/02
Flint River Reservoir	Intake for Smith WTP, GA	9/6/02

*WWTP = conventional municipal wastewater treatment plant; WTP = water treatment plant; AS = Activated sludge; UV = ultraviolet disinfection; Cl_2 = chlorine disinfection; HRT = hydraulic retention time; MWD = Metropolitan (CA) Water District.

is equipped with primary screening and clarification followed by pure oxygen activated sludge treatment, clarification and chlorine disinfection. The samples analyzed as part of the occurrence survey were collected at the West Basin AWWTP, which treats the secondary effluent from the Hyperion treatment plant.

Mt. View Wastewater Treatment Plant (Martinez, CA)

The Mt. View municipal wastewater treatment plant is a 0.06 $m^3\ s^{-1}$ (1.5 MGD) facility equipped with primary screening and clarification followed by a trickling filter for secondary treatment and a biotower for ammonia removal. The effluent is subjected to ultraviolet disinfection prior to being discharged to an engineered treatment wetland.

Roger Road Wastewater Treatment Plant (Tuscon, AZ)

The Roger Road municipal wastewater treatment plant is a 1.4 m^3 s^{-1} (31 MGD) facility equipped with primary screening and clarification, followed by activated sludge treatment and chlorine disinfection. The treatment plant discharges directly to an infiltration pond that recharges an aquifer at the Sweetwater recharge facility. During the occurrence survey, samples collected from the infiltration pond were assumed to be representative of the effluent from Roger Road treatment plant.

San Jose/Santa Clara Water Pollution Control Plant (San Jose, CA)

The San Jose municipal wastewater treatment plant is a 7.3 m^3 s^{-1} (167 MGD) facility equipped with primary screening and clarification, followed by activated sludge treatment with biological nutrient removal in activated sludge reactors. Following clarification, the wastewater undergoes dual media filtration and chlorine disinfection.

South Cobb Wastewater Treatment Plant (Cobb County, GA)

The South Cobb municipal wastewater treatment plant is a 1.8 m^3 s^{-1} (40 MGD) facility equipped with primary treatment and aerated activated sludge treatment followed by chlorine disinfection.

Wastewater Treatment Plant S (location name withheld at request of utility)

The S municipal wastewater treatment plant is a 6.6 m^3 s^{-1} (150 MGD) facility equipped with primary screening and clarification, followed by pure oxygen activated sludge treatment and chlorine disinfection. The operators of the treatment plant requested that the plant's identity remain anonymous and it is designated as treatment plant S throughout this report.

F. Wayne Hill Water Resources Center (Gwinnett County, GA)

The F. Wayne Hill facility is a 0.88 m^3 s^{-1} (20 MGD) advanced wastewater treatment plant. The facility consists of primary treatment followed by activated sludge treatment in a reactor operated for biological nutrient removal. After the secondary clarification, the wastewater undergoes ferric chloride addition followed by dual-media filtration, pre-ozonation, granular activated carbon filtration and ozonation.

West Basin Municipal Water Advanced Wastewater Treatment Plant (Los Angeles, CA)

The West Basin treatment plant is an advanced wastewater treatment plant consisting of two treatment trains. The first treatment train uses lime coagulation followed by cellulose acetate membranes while the second train uses microfiltration followed by reverse osmosis with thin-film composite membranes. The two trains are combined prior to ultraviolet disinfection. As part of the occurrence survey samples were collected from the second treatment train.

Advanced Treatment Plant T (location name withheld at request of utility)

Treatment plant T is an advanced wastewater treatment plant that is part of an indirect potable water reuse system. The treatment plant consists of two treatment trains that treat wastewater effluent from a municipal wastewater treatment plant. During the occurrence survey, samples were collected from the treatment train that consists of microfiltration, reverse osmosis with thin-film composite membranes and ultraviolet disinfection in the presence of hydrogen peroxide.

Sweetwater Recharge Facility (Tuscon AZ)

The Sweetwater groundwater recharge site consists of an infiltration pond that receives wastewater effluent from the Roger Road wastewater treatment plant. The underlying aquifer is equipped with an extensive network of monitoring wells. As a part of the occurrence survey, groundwater was collected from two downgradient wells: (1) a shallow well screened at 5.1 meters; and, (2) a deep well screened at approximately 30.5 m. According to tracer data collected at the site the water sampled from both wells consists entirely of wastewater effluent (i.e., there is no dilution with local groundwater) and has a residence time in the aquifer of approximately 2.5 and 15 days, respectively.

Mt. View Engineered Treatment Wetland (Martinez, CA)

The Mt. View engineered treatment wetlands consist of a series of five ponds in series connected by weirs and underground piping. The wetland ponds are approximately 1.5 meters deep and are extensively vegetated along the edges with cattails duckweed. The mean hydraulic residence time of the wetland is approximately 7 days.

Prado Engineered Treatment Wetlands (Orange County, CA)

The Prado Engineered Treatment wetlands treat water from the Santa Ana River. During summertime, most of the water in the Santa Ana River originates at Riverside and San Bernardino tertiary wastewater treatment plants located approximately 20 km upstream. During other times of the year, the river receives a combination of stormwater runoff and wastewater discharge from the upstream watershed. The wetland consists of a series of treatment cells vegetated with cattail and duckweed.

SAMPLE COLLECTION AND ANALYSIS

After completion of the method development activities described in the previous chapter, samples from each of the sites listed in Table 3.1 were analyzed for the target PhACs. In most cases, samples were collected on more than one occasion. Results of the analyses are summarized in Table 3.2-3.4 and Figures 3.1 through 3.11. In some cases, the results for one or more analyte did not meet our QA/QC criteria. In these cases, a data qualifier is included in the table and the results are not included in any of the summary statistics discussed later in this chapter.

Results for the acidic drugs are summarized in Table 3.2. Each of the acidic drugs was detected in at least one effluent samples collected from a conventional wastewater treatment plant (Figure 3.1). With the exception of one effluent sample from the secondary effluent that is received by treatment plant T, gemfibrozil and naproxen were present at the highest concentrations. Diclofenac was detected in most of the effluent samples more frequently than ibuprofen, indometacine and ketoprofen. On four different dates, samples also were collected before chlorine disinfection to assess the potential transformation of PhACs during chlorine disinfection. This issue is discussed in more detail later in this chapter. Results of several preliminary experiments indicated that diclofenac, gemfibrozil, indometacine and naproxen react with chlorine. For two of the wastewater effluent samples recoveries of the internal standard, mecoprop, were above our target value of 120%. We are confident that acidic drugs were present in these samples, but due to the high recoveries of the internal standard, the accuracy of data is questionable.

Acidic drugs were not detected in any of the final effluent samples from the advanced wastewater treatment systems (Table 3.2, Figure 3.2). It should be noted that the internal standard was not recovered in the reverse osmosis effluent sample from treatment plant T on October 16, 2002. This could be attributable to failure to add the internal standard to the sample or a matrix recovery problem. The recovery of the internal standard in the sample from the West Basin advanced WWTP, one of the other two reverse osmosis effluent samples, was slightly higher than our target value (i.e., 126% vs. an upper target value of 120%). Despite these shortcomings and the limited number of samples, it is evident that reverse osmosis lowers the concentrations of acidic drugs to values below the limit of quantification. The good removal of acidic drugs is consistent with expectations for charged organic compounds with molecular weights in this range. Concentrations of acidic drugs did not decrease appreciably during microfiltration, which is consistent with expectations for this process, which mainly removes micron-sized particles. Acidic drugs also were not detected in the F. Wayne Hill advanced WWTP after granular activated carbon treatment. However, the recovery of C_{13} mecoprop was below the target values in these two samples, and confirmatory analysis may be needed to verify the removal of these compounds during treatment.

At the Sweetwater soil aquifer treatment system (Table 3.2, Figure 3.3), concentrations of acidic drugs decreased between the pond and the shallow well. Complete removal of acidic drugs was observed in the deep well.

The fate of acidic PhACs in engineered treatment wetlands is unclear. Gemfibrozil and naproxen were detected in the effluent discharged into the Mt. View engineered treatment wetland (Figure 3.4). The concentration of gemfibrozil was approximately equal in samples collected from the beginning, middle and end of the wetland system while the concentration of naproxen appeared to decrease in the wetland. However, these data should be interpreted with caution because of the limited number of observations and the variation in the concentrations of PhACs in the wastewater effluent could be responsible for these apparent trends. With the exception of one sample, acidic drugs were not detected in the Prado engineered treatment wetlands. This could have been attributable to the discharge of snowmelt or groundwater to the river during this period.

Table 3.2

Concentrations of acidic drugs measured during the occurrence survey

Location	Date	Concentration* (ng/L)						Recovery (%)
		Diclofenac	Gemfibrozil	Ibuprofen	Indometacine	Ketoprofen	Naproxen	Mecoprop
Effluent from conventional wastewater treatment plants								
Dublin/San Ramon[‡]	3/21/02	<10	2500	130	41	<10	1700	117%
Dublin/San Ramon	3/21/02	<10	1700	170	35	<10	1400	80%
Hyperion	9/18/01	59	1300	91	<10	55	250	64%
F. Wayne Hill (2° effluent)	10/17/02	60	92	<10	36	<10	190	77%
Mt View	9/4/01	<10[†]	300[†]	<10[†]	200[†]	<10[†]	370[†]	168[†]%
Roger Road	4/1/02	41[†]	1500[†]	150[†]	66[†]	<20[†]	200[†]	6[†]%
Plant T (2° effluent)	9/12/01	200[†]	4600[†]	2300[†]	<10[†]	47[†]	1800[†]	180[†]%
Plant T (2° effluent)	10/16/02	78	2000	160	33	32	350	79%
San Jose/Santa Clara[‡]	3/26/02	43	270	<10	20	<10	220	77%
San Jose/Santa Clara	3/26/02	44	310	<10	21	<10	200	67%
San Jose/Santa Clara[‡]	6/26/02	60	250	<10	<10	11	230	66%
San Jose/Santa Clara	6/26/02	35	240	<10	<10	<10	100	72%
Treatment Plant S[‡]	7/1/02	48	640	<10	<10	14	290	98%
Treatment Plant S	7/1/02	60	540	<10	<10	<10	640	101%
Treatment Plant S	10/4/02	62	5500	320	<10	<10	3200	77%
Effluent from advanced wastewater treatment plants								
F. Wayne Hill GAC Effluent	10/17/02	<10[†]	<10[†]	<10[†]	<10[†]	<10[†]	<10[†]	37[†]%
F. Wayne Hill Final Effluent	10/17/02	<10[†]	<10[†]	<10[†]	<10[†]	<10[†]	<10[†]	11[†]%
Plant T RO Effluent	9/12/01	<10	<10	<10	<10	<10	<10	98%
Plant T MF Effluent	10/16/02	77	1700	160	42	35	510	85%
Plant T RO Effluent	10/16/02	<10[†]	<10[†]	<10[†]	<10[†]	<10[†]	<10[†]	0[†]%
West Basin MF Effluent	9/18/01	110[†]	2300[†]	<10[†]	26[†]	72[†]	300[†]	161[†]%
West Basin RO Effluent	9/18/01	<10[†]	<10[†]	<10[†]	<10[†]	<10[†]	<10[†]	126[†]%

(continued)

Table 3.2 (Continued)

Location	Date	Concentration* (ng/L)						Recovery (%)
		Diclofenac	Gemfibrozil	Ibuprofen	Indometacine	Ketoprofen	Naproxen	Mecoprop
Sweetwater soil aquifer treatment system								
Shallow Well	4/1/02	36	590	220	56	<20	480	67%
Deep Well	4/1/02	<20	<20	<10	<10	<20	<10	64%
Engineered Treatment Wetlands								
Mt View beginning	9/4/01	<10†	120†	20†	<10†	<10†	200†	122†%
Mt View middle	9/4/01	<10	92	<10	<10	<10	19	113%
Mt View end	9/4/01	<10	110	<10	<10	<10	28	109%
Prado beginning	4/6/02	<20	<20	<10	<10	<20	<10	68%
Prado middle	4/6/02	<20†	22†	<10†	<10†	<20†	<10†	33†%
Prado end	4/6/02	<20†	<20†	<10†	<10†	<20†	<10†	59†%

*Results are tabulated with two significant figures and are not corrected for recovery of the internal standard (mecoprop).

†Recovery of internal standard outside of the acceptable recovery range.

‡Sample collected immediately prior to effluent chlorination.

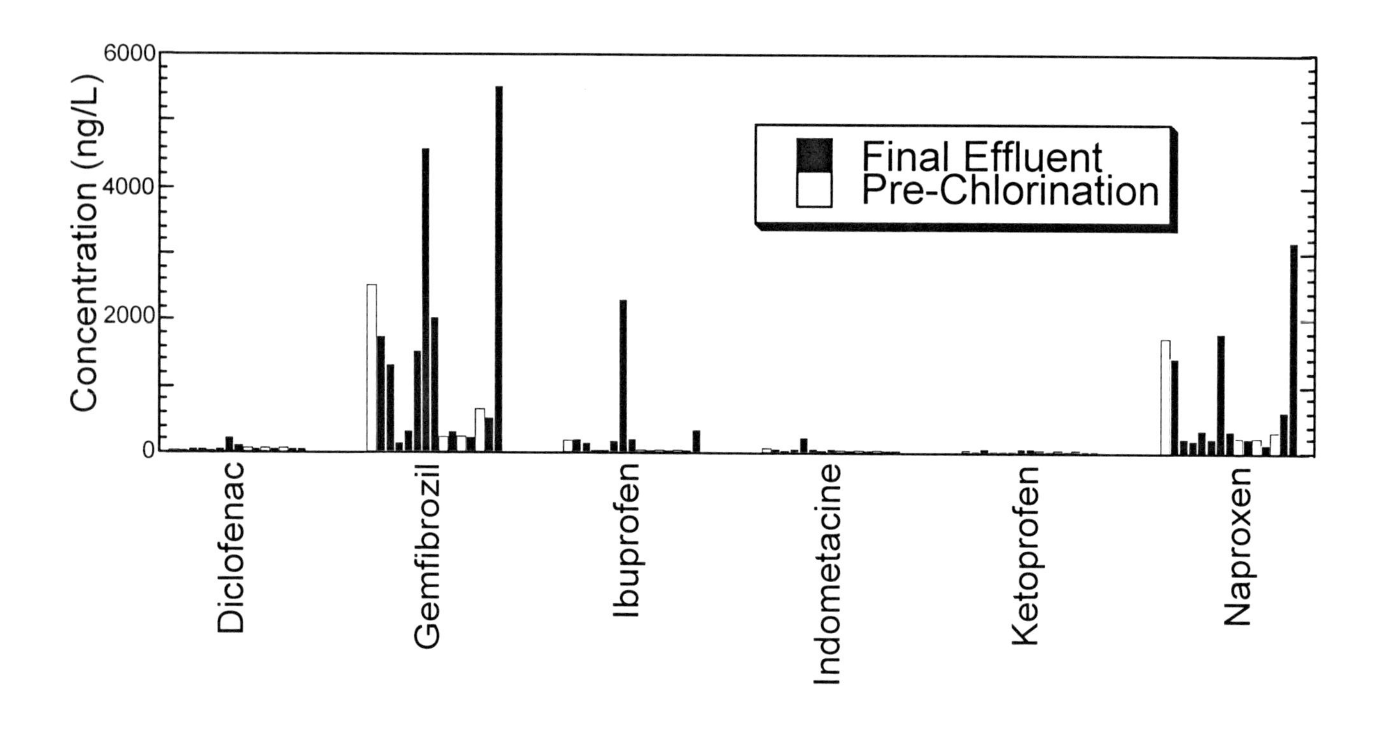

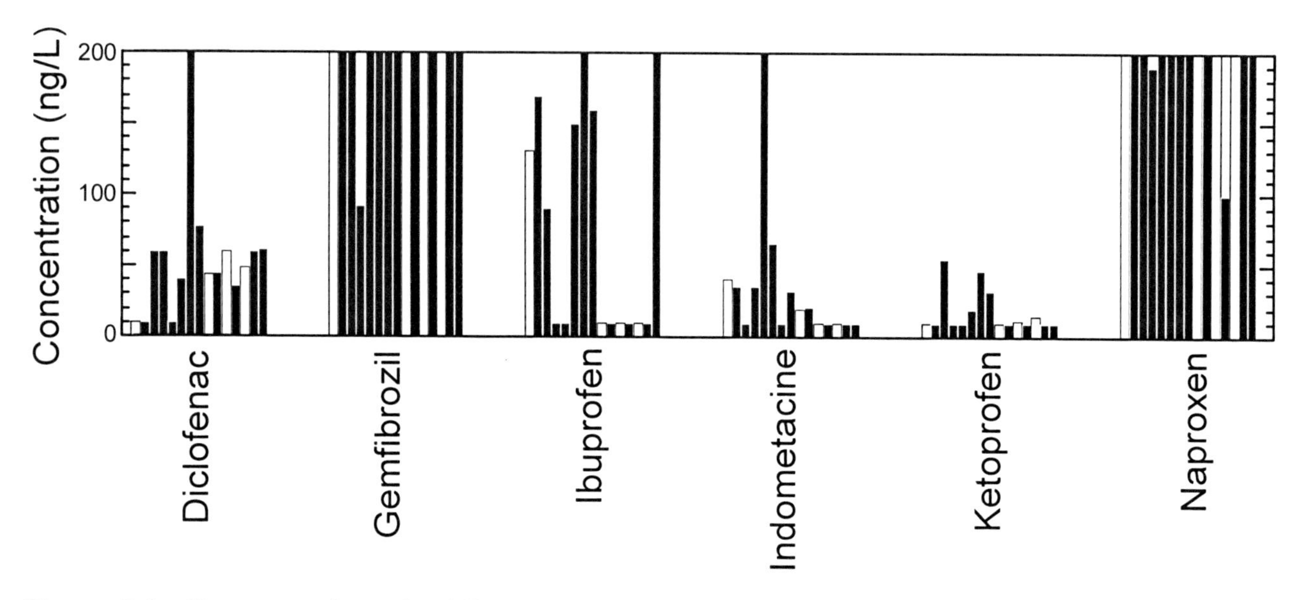

Figure 3.1 Concentration of acidic drugs in effluent samples collected from conventional WWTPs. Lower figure depicts smaller different y-axis range. Concentrations below detection limit are depicted at the detection limit.

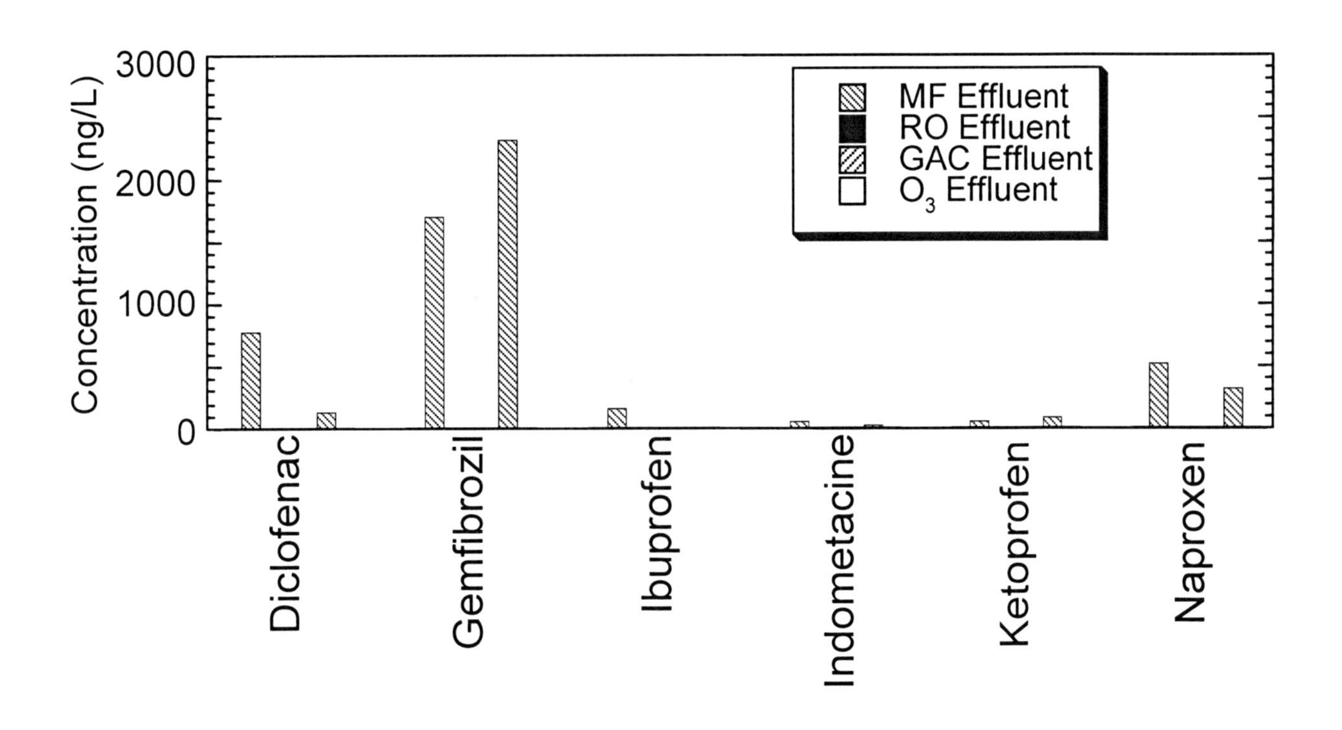

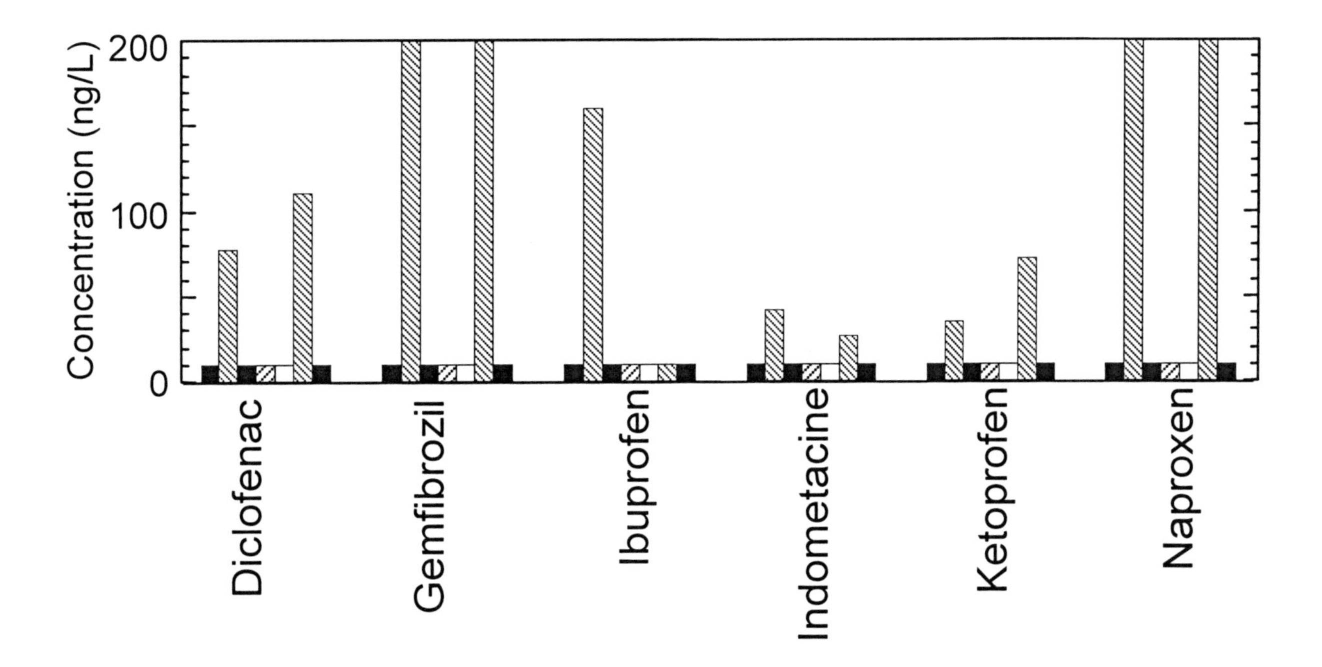

Figure 3.2 Concentration of acidic drugs in effluent samples collected from advanced WWTPs. Lower figure depicts smaller different y-axis range. Concentrations below detection limit are depicted at the detection limit (10 ng/L).

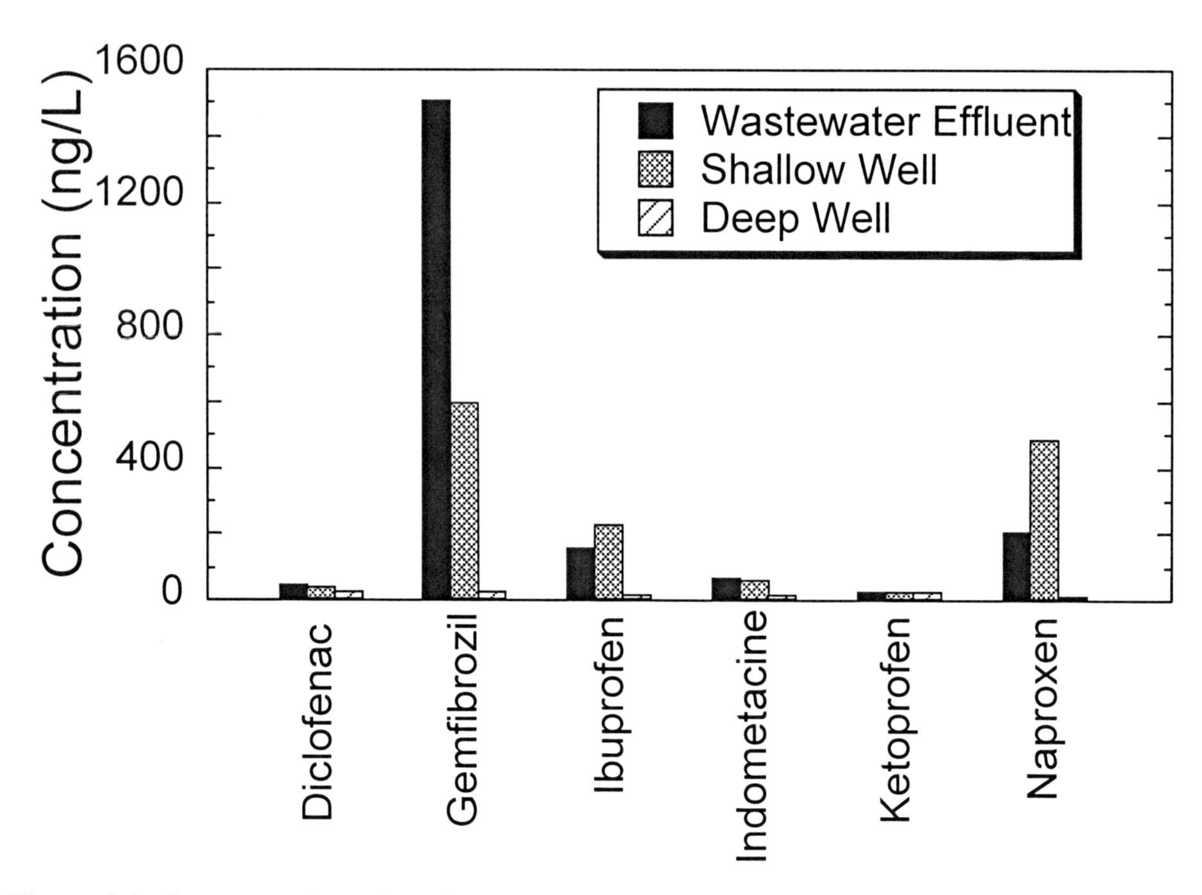

Figure 3.3 Concentration of acidic drugs in effluent samples collected from the Sweetwater soil aquifer treatment system. Concentrations below detection limit are depicted at the detection limit.

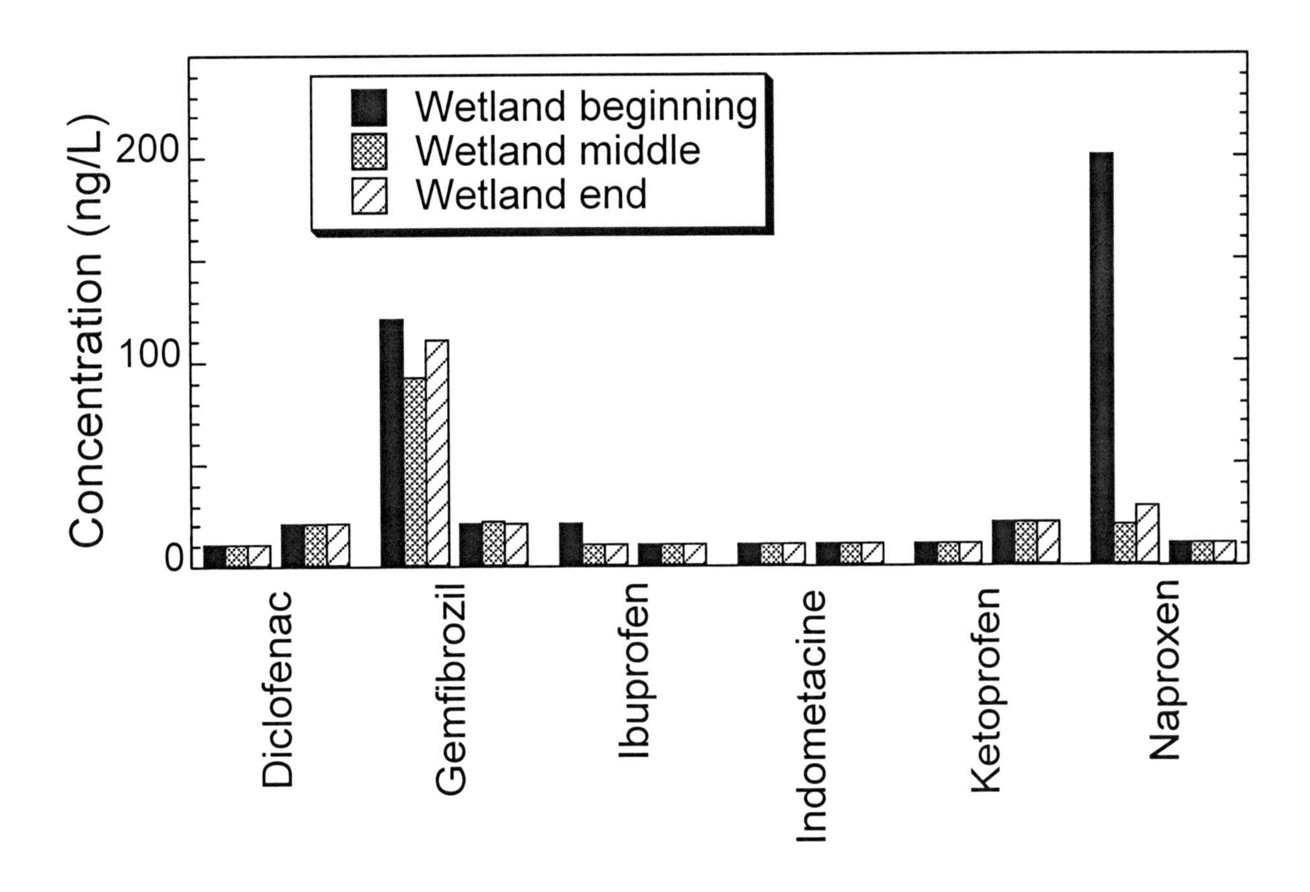

Figure 3.4 Concentration of acidic drugs in effluent samples collected from engineered treatment wetlands. First set of data indicates Mt. View wetland, second set of data indicate Prado wetlands. Concentrations below detection limit are depicted at the detection limit.

Both of the beta-blockers (metoprolol and propranolol) were detected frequently in municipal wastewater effluent (Table 3.3, Figure 3.5). Metoprolol was detected in all but one sample while propranolol was detected in 9 of the 14 wastewater effluent samples. The concentration of metoprolol was usually higher than that of propranolol and no clear pattern was evident in the concentrations among the different treatment plants. The recovery of both compounds in spiked wastewater effluent samples ranged between 47% and 76%.

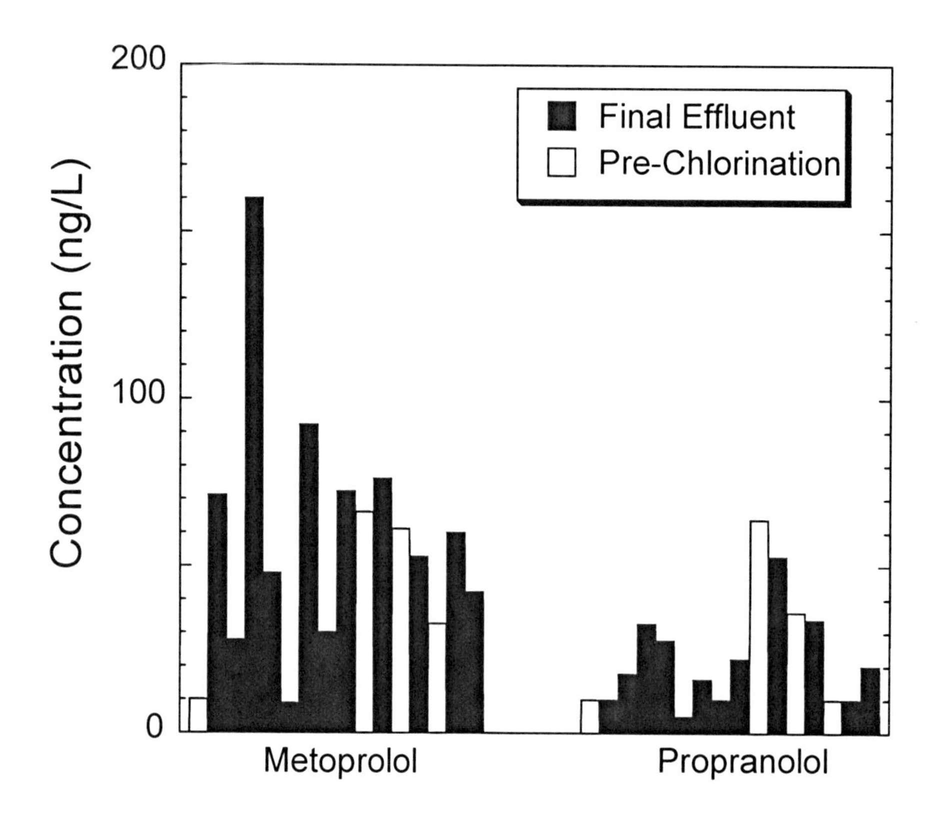

Figure 3.5 Concentration of beta-blockers in effluent samples collected from conventional WWTPs. Concentrations below detection limit are depicted at the detection limit.

Table 3.3

Concentrations of beta-blockers measured during the occurrence survey

Location	Date	Metoprolol		Propranolol	
		Conc'n (ng/L)	Recovery	Conc'n (ng/L)	Recovery
Effluent from conventional wastewater treatment plants					
Dublin/San Ramon*	3/21/02	<10	76%	<10	65%
Dublin/San Ramon	3/21/02	71	76%	<10	65%
F. Wayne Hill (2° eff.)	10/17/02	28		18	
Hyperion	9/18/01	160	48%	33	47%
Mt View	9/4/01	48	53%	28	62%
Mt View	8/21/02	9	55%	5	70%
Roger Road	4/1/02	92	67%	16	70%
Plant T Influent	9/12/01	30	55%	<10	45%
Plant T Influent	10/16/02	72		22	
San Jose/Santa Clara*	3/26/02	66	76%	64	65%
San Jose/Santa Clara	3/26/02	76	76%	53	65%
San Jose/Santa Clara*	6/26/02	61	70%	36	54%
San Jose/Santa Clara	6/26/02	53	70%	34	54%
Treatment Plant S*	7/1/02	33	70%	<10	54%
Treatment Plant S	7/1/02	60	70%	<10	54%
Treatment Plant S	8/14/02	42	70%	20	54%
Effluent from advanced wastewater treatment plants					
F. Wayne Hill GAC Eff.	10/17/02	<10		<10	
F. Wayne Hill Final Eff.	10/17/02	<10		<10	
Plant T MF Effluent	9/12/01	12	55%	<10	45%
Plant T RO Effluent	9/12/01	23	46%	36	35%
Plant T MF Effluent	10/16/02	130		33	
Plant T RO Effluent	10/16/02	<10		<10	
West Basin MF Effluent	9/18/01	67	48%	61	47%
West Basin RO Effluent	9/18/01	<10	48%	<10	47%
Sweetwater soil aquifer treatment system					
Shallow Well	4/1/02	23	67%	13	70%
Deep Well	4/1/02	14	67%	13	70%
Engineered Treatment Wetlands					
Mt View middle	9/4/01	<10	39%	<10	62%
Mt View end	9/4/01	<10	7%	<10	34%
Prado beginning	4/6/02	17	60%	12	54%
Prado middle	4/6/02	13	60%	11	54%
Prado end	4/6/02	16	60%	12	54%

*Sample collected immediately prior to effluent chlorination.

Analysis of samples from advanced treatment plant T indicated the presence of metoprolol and propranolol in one of the two samples collected after reverse osmosis treatment while samples from a similar treatment system at the West Basin advanced wastewater treatment plant indicated complete removal of the beta-blockers (Table 3.3, Figure 3.6). Matrix spike recoveries for samples from the advanced wastewater treatment plants were comparable to those measured in the effluent samples from the conventional WWTPs. In both cases the concentrations of the compounds were near the limit of quantification (10 ng/L). The detection of beta-blockers in the reverse osmosis permeate at treatment plant T does not appear to be an artifact related to matrix interference because the chromatograms were free from noise at the retention times of the compounds and the blanks did not show contamination. The apparent discrepancy between these results could be related to some aspect of the membrane performance or differences between operating conditions at the two treatment plants. However, additional sampling would be required to confirm the breakthrough of the beta-blockers in reverse osmosis systems.

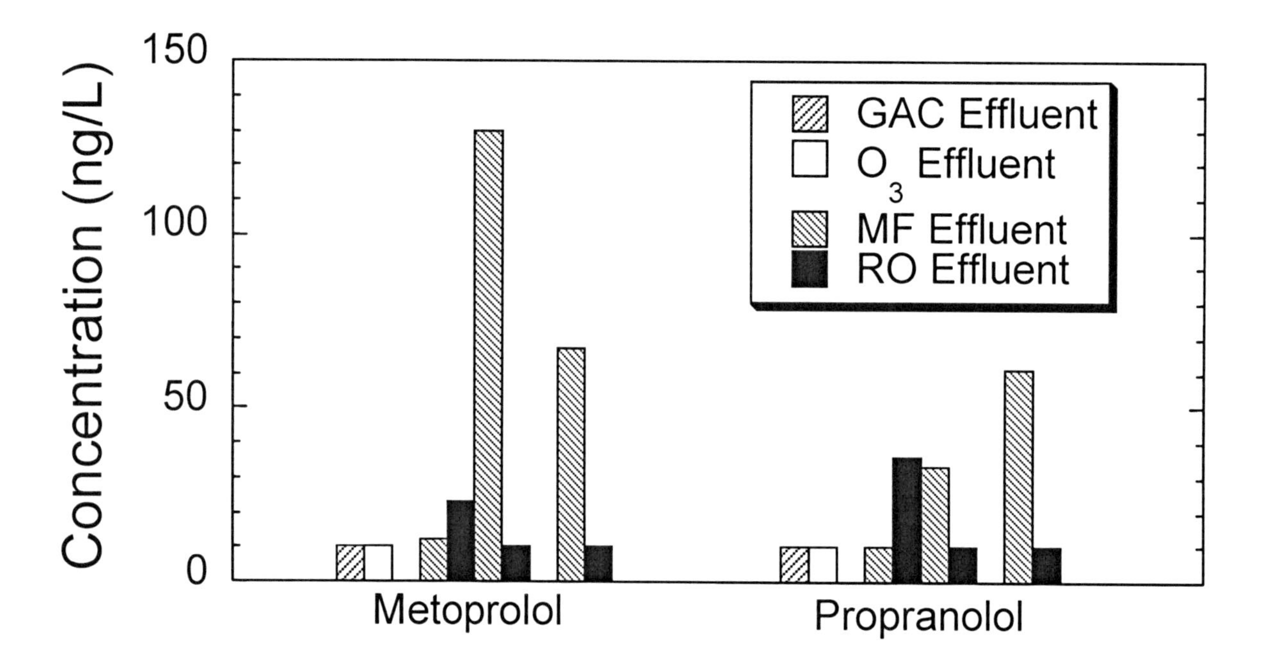

Figure 3.6 Concentration of beta-blockers in effluent samples collected from advanced WWTPs. Concentrations below detection limit are depicted at the detection limit.

The beta-blockers also were detected in samples collected from the Sweetwater soil aquifer treatment site and the Prado engineered treatment wetlands (Figure 3.7). At the Sweetwater soil aquifer treatment system, the concentration of both compounds decreased between as the wastewater effluent from the Roger Road WWTP infiltrated to the first well (Table 3.3, Figure 3.7). The beta-blockers were detected at the Prado engineered treatment wetland at concentrations slightly above the limit of quantification. These results suggest that beta-blockers are more resistant to attenuation in the receiving waters than the acidic drugs.

At least one antibiotic was detected in each of the final effluent samples collected from conventional WWTPs (Table 3.4, Figure 3.8). The antibiotics sulfamethoxazole and trimethoprim were detected at the highest concentrations, followed by ofloxacin and ciprofloxacin. Enrofloxacin, norfloxacin and sulfamethazine were detected in only one or two of the effluent samples. Concentrations of ciprofloxacin and trimethoprim were lower after chlorine disinfection. However, variability in concentrations between grab samples could have accounted for some of the apparent decreases. Matrix interference prevented quantification of antibiotics in several samples (e.g., sulfamethazine and sulfamethoxazole in Hyperion WWTP effluent) and matrix effects led to unacceptable recoveries in other wastewater effluent samples.

Antibiotics were removed in advanced wastewater treatment plants (Table 3.4, Figure 3.9). Concentrations of antibiotics were below the limit of quantification in all of the reverse osmosis permeate samples collected from the West Basin advanced WWTP. All of the antibiotics other than enrofloxacin were detected in one of the two GAC effluent samples (i.e., 4/22/02) collected from the F. Wayne Hill advanced wastewater treatment plant. The other GAC

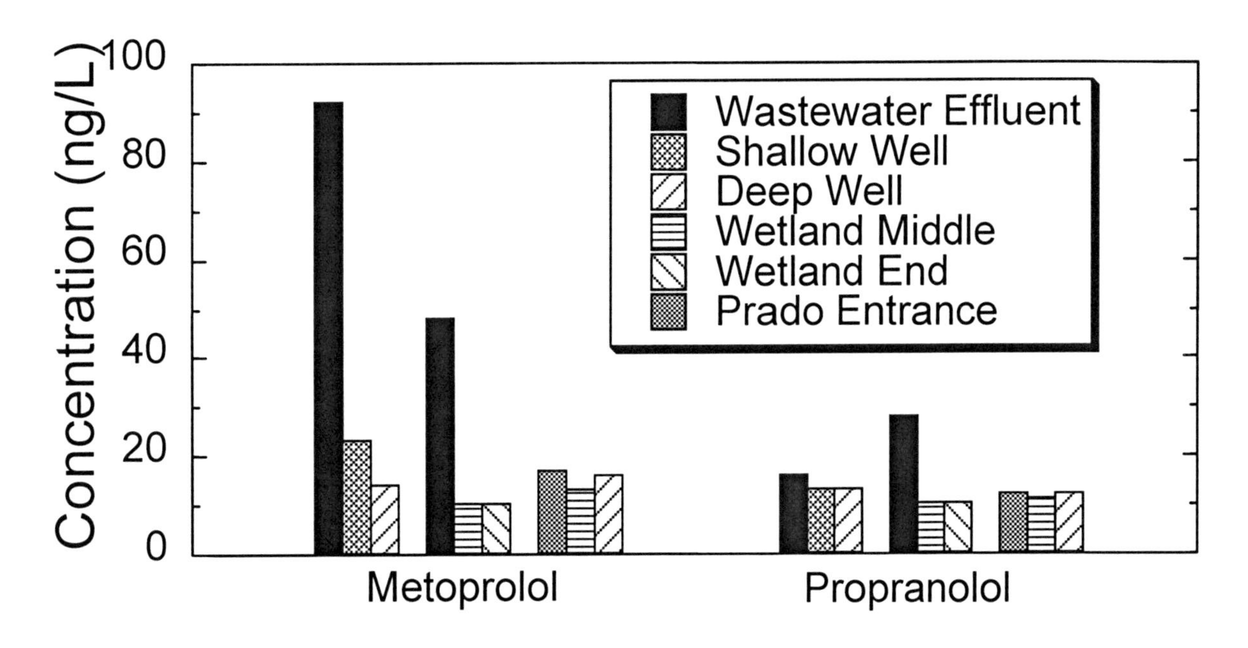

Figure 3.7 Concentration of beta-blockers in effluent samples collected from soil aquifer treatment systems and engineered treatment wetlands. Data from the Mt. View treatment wetland are displayed before data from the Prado treatment wetland. Concentrations below detection limit are depicted at the detection limit.

Table 3.4
Concentrations of antibiotics measured during the occurrence survey

Location	Date	Concentration* (ng/L) Ciprofloxacin	Enrofloxacin	Norfloxacin	Ofloxacin	Sulfamethazine	Sulfameth-oxazole	Trimethoprim
Effluent from conventional wastewater treatment plants								
Hyperion	5/22/02	350	<60	<60	450	NA	NA	690
Hyperion	9/10/02	310[†]	<60[†]	<60[†]	490[†]	NA	NA	400[†]
F. Wayne Hill (2° effluent)	4/22/02	190[†]	<60[†]	190	130[†]	500	2000	1900
F. Wayne Hill (2° effluent)	7/17/02	<60[†]	<60[†]	<60[†]	1100[†]	<60[†]	200[†]	<60[†]
Mt View	4/9/02	510	150	80	300[†]	300[†]	1500	550
Mt View	10/15/02	860	<60	<60	180	<60[†]	1400	460
Roger Road	2/17/02	<60	<60	<60	<60[†]	<60	NA	470
Roger Road	4/1/02	270	140	<60	600	NA	NA	750[†]
South Cobb[‡]	2/15/02	100	<60	<60	<60	<60	<60	1210
South Cobb	2/15/02	<30	<30	<30	<30	<30	<30	1210
South Cobb[‡]	6/12/02	<60	<60	<60[†]	300	<60	580	<60
South Cobb	6/12/02	<30[†]	<30[†]	<30[†]	45	<30	60[†]	<30
Clayton[‡]	3/19/02	270	<60	<60	260	<60[†]	1500	700
Clayton	7/1/02	<30	<30	<30	200	<30	2000	380
Clayton[‡]	8/20/02	100	<60	<60	140	NA	210	750
Clayton	8/20/02	<30	<30	<30	130	NA	385	<30[†]
Effluent from advanced wastewater treatment plants								
F. Wayne Hill GAC Effluent	4/22/02	180	<30	70	40	450	80	1500
F. Wayne Hill Final Effluent	4/22/02	<30[†]	<30[†]	<30[†]	<30[†]	110	<30	<30[†]
F. Wayne Hill GAC Effluent	7/17/02	<30[†]	<30[†]	<30[†]	<30[†]	<30[†]	<30[†]	<30
F. Wayne Hill Final Effluent	7/17/02	<30	<30	<30[†]	<30[†]	<30	<30	<30
West Basin MF Effluent	5/22/02	<60	<60	<60[†]	560	NA	NA	670

(continued)

Table 3.4 (Continued)

Location	Date	Concentration* (ng/L)						
		Ciprofloxacin	Enrofloxacin	Norfloxacin	Ofloxacin	Sulfamethazine	Sulfameth-oxazole	Trimethoprim
Effluent from conventional wastewater treatment plants								
West Basin RO Effluent	5/22/02	<30†	<30†	<30†	<30†	<30†	<30†	<30†
West Basin MF Effluent	9/10/02	140†	<60	<60†	190	NA	1000	550
West Basin RO Effluent	9/10/02	<30	<30	<30	<30	<30	<30	<30
Sweetwater soil aquifer treatment system								
Shallow Well	2/17/02	<60	<60	<60	<60†	<60†	NA	470
Deep Well	2/17/02	<60	<60	<60	<60	<60†	NA	<60
Shallow Well	4/1/02	270	<60	<60	<60	NA	540	110†
Deep Well	4/1/02	<30	<30	<30	<30	<30	80	35
Engineered Treatment Wetlands								
Mt View beginning	4/9/02	<60	<60	<60	<60	<60	270	140†
Mt View end	4/9/02	120	<60	<60	<60†	<60†	340	70†
Mt View beginning	10/15/02	<60	<60	<60	260	<60†	1000†	90
Mt View end	10/15/02	<60	<60	<60	120	<60†	1000†	<60

*Results are tabulated with two significant figures. For samples collected before 4/1/02 the internal standard method was used for quantification. For samples collected after 4/1/02 the standard addition method was used for quantification.
†Recovery of internal standard outside of the acceptable recovery range.
‡Sample collected immediately prior to effluent chlorination or prior to ultraviolet disinfection.

effluent sample from the F. Wayne Hill plant did not contain detectable concentrations of antibiotics. However, secondary effluent samples collected from the plant on that date (i.e., 7/17/02) contained lower concentrations of antibiotics and unacceptable recoveries were obtained in all of the samples from that date. Therefore, it is impossible to determine if the breakthrough of antibiotics observed on 4/22/02 is a unique event. Among the antibiotics detected in the GAC effluent on 4/22/02, only sulfamethazine was detected after ozonation. However, the plant operators report that the ozonation system was not operating correctly during this period.

Analysis of samples from the Sweetwater soil aquifer treatment system indicated that removal occurs as samples pass through the subsurface, but that removal is not always complete (Table 3.4, Figure 3.10). Samples were analyzed for antibiotics on two different dates. On the first date (2/17/02) trimethoprim was the only antibiotic detected in the effluent from the wastewater treatment plant that discharged to the system. The concentration of trimethoprim in the shallow well was equal to that detected in the effluent sample and no trimethoprim was detected in the deep groundwater sample. On the second sampling date (4/1/02), ciprofloxacin, enrofloxacin, ofloxacin and trimethoprim were detected in the wastewater effluent sample. Sulfamethoxazine and sulfamethoxazole were not analyzed in the effluent sample due to matrix interference. Ciprofloxacin, sulfamethoxazole and trimethprim were detected in the shallow well while sulfamethoxazole and trimethoprim were detected in the deep well.

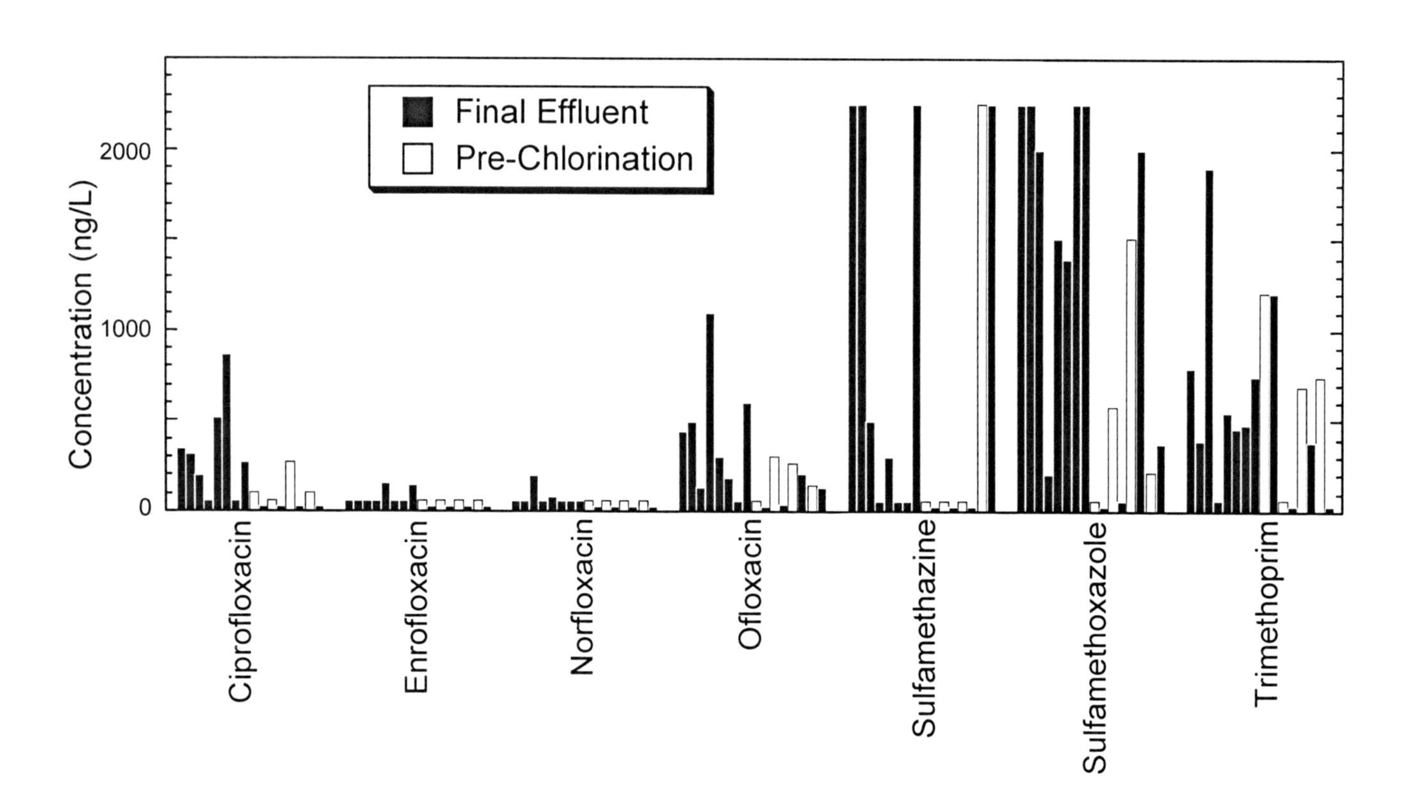

Figure 3.8 Concentration of antibiotics in effluent samples collected from conventional WWTPs. Concentrations below detection limit are depicted at the detection limit.

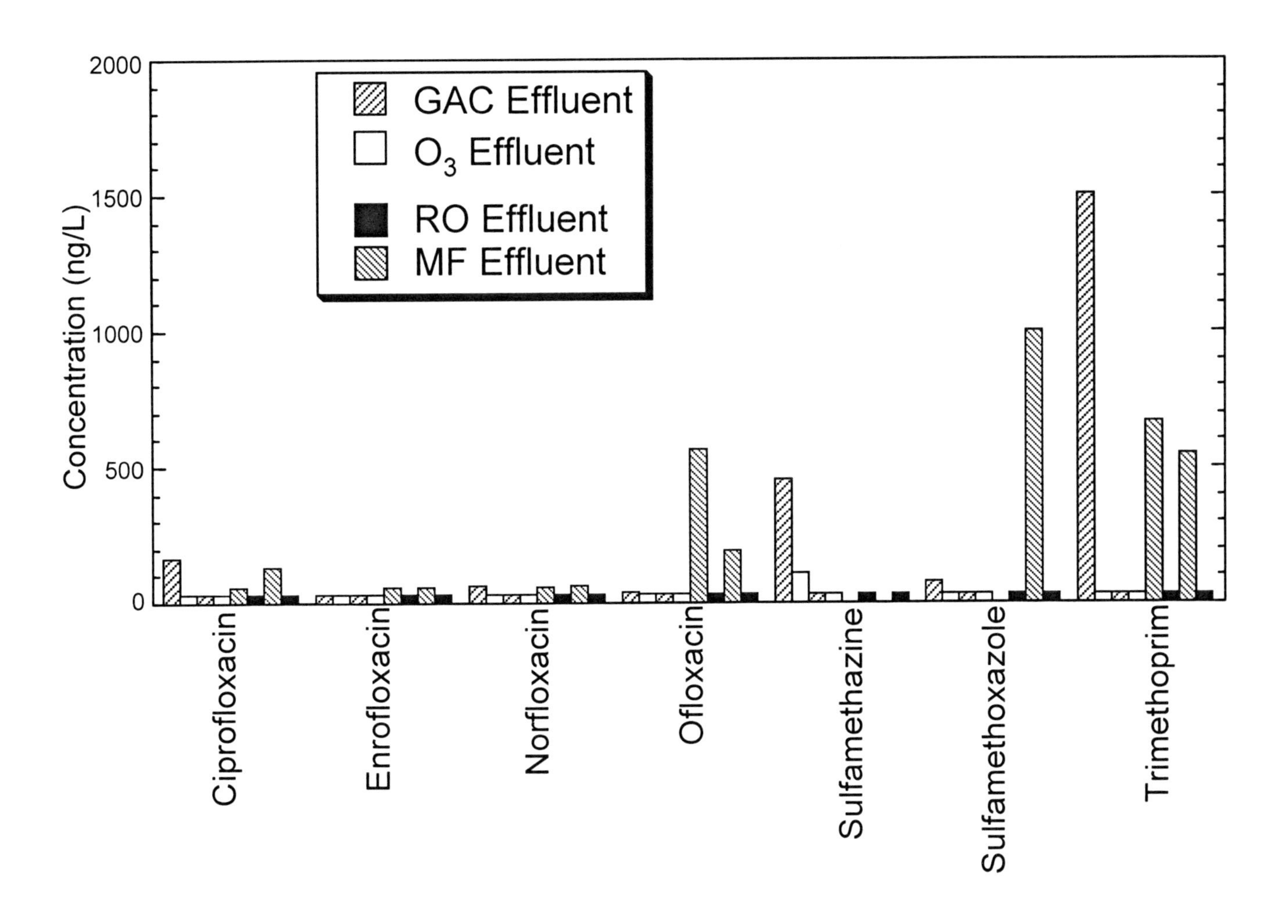

Figure 3.9 Concentration of antibiotics in effluent samples collected from advanced WWTPs. Concentrations below detection limit are depicted at the detection limit.

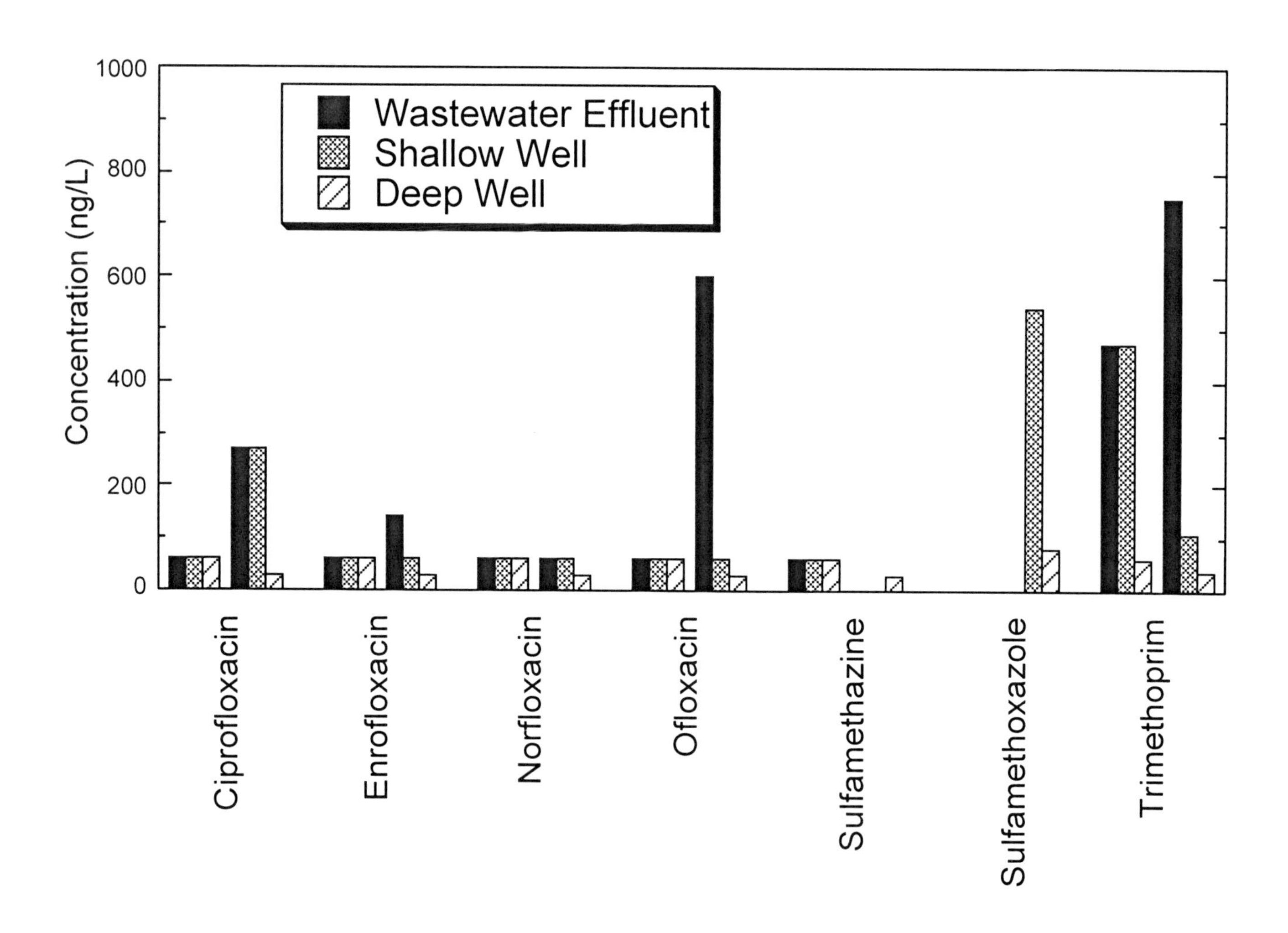

Figure 3.10 Concentration of antibiotics in samples collected from the Sweetwater soil aquifer treatment system. Concentrations below detection limit are depicted at the detection limit.

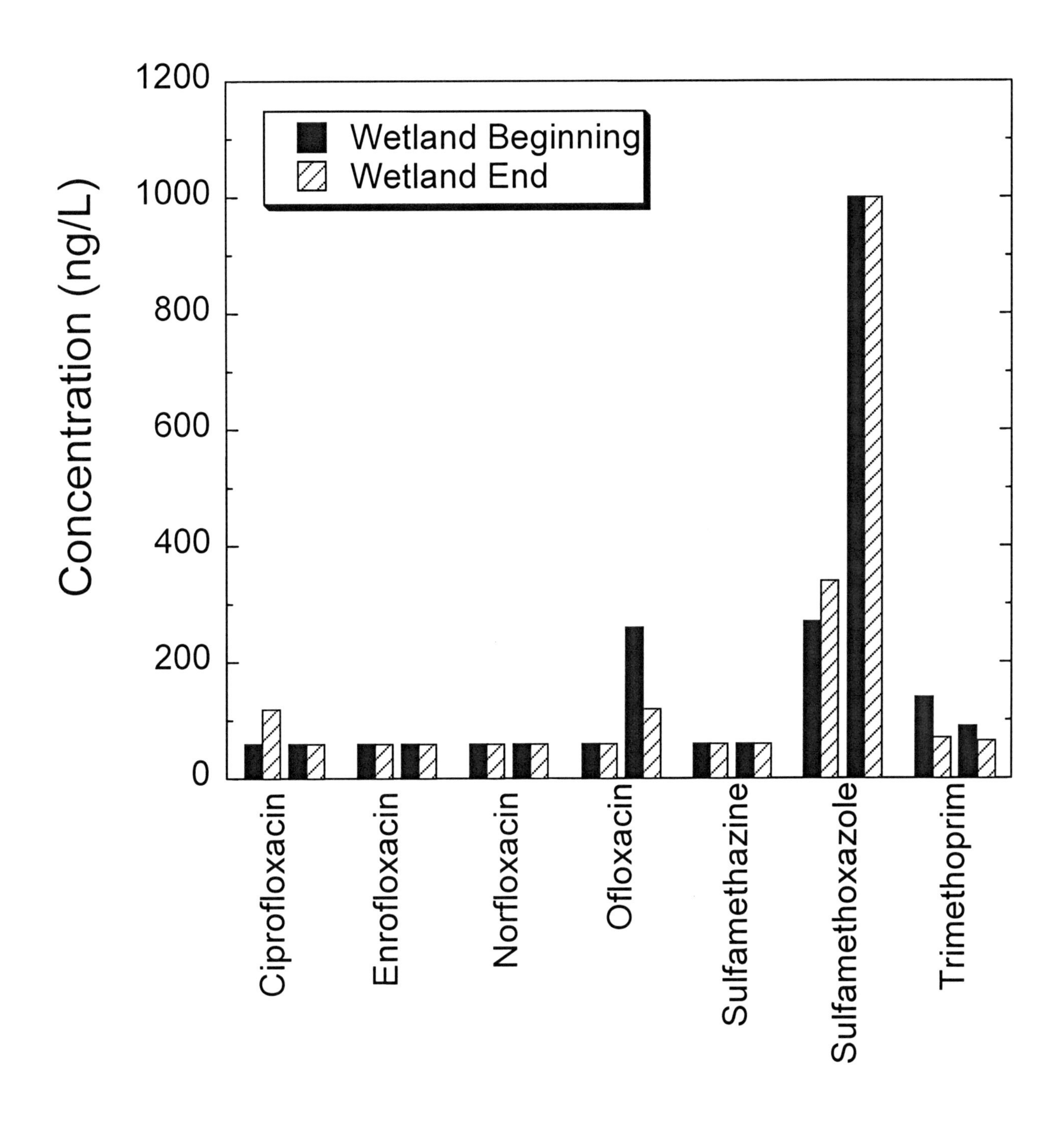

Figure 3.11 Concentration of antibiotics in samples collected from the Mt. View engineered treatment wetland. Concentrations below detection limit are depicted at the detection limit.

As mentioned previously, results from preliminary experiments indicated that some of the PhACs could be transformed during chlorine disinfection. Analysis of acidic drugs and beta-blockers before and after chlorine disinfection (Table 3.2, 3.3) did not provide any clear evidence of transformation during chlorine disinfection at the Dublin/San Ramon, San Jose/Santa Clara and S WWTPs. However, it would be difficult to detect a modest decrease in concentration by collection of samples before and after a unit process, especially when the uncertainty of the analytical measurements is relatively large (e.g., analysis of duplicate samples often indicates a coefficient of variation of approximately 20%).

To further investigate the potential for removal of PhACs during chlorine disinfection, samples were collected from the San Jose/Santa Clara WWTP and Treatment Plant S. To gain further insight into the potential importance of reactions that occur during chlorine disinfection, we conducted an experiment using wastewater effluent collected before disinfection at the San Jose/Santa Clara WWTP and Treatment Plant S. As part of this experiment, we added 1,000 ng/L of each of the target analytes to the wastewater prior to addition of a low dose (i.e., 0.14 mM or 10 mg/L as Cl_2) and a high dose (i.e., 0.86 mM or 60 mg/L as Cl_2) of chlorine. After one hour, the chlorine was quenched by addition of an excess of sodium thiosulfate.

Results from the experiment (Figure 3.12) provide evidence that chlorine disinfection removes some of the pharmaceuticals from wastewater that does not contain high concentrations of ammonia. Upon exposure to free chlorine, diclofenac and naproxen were almost completely removed by both the low and high concentrations of free chlorine. Ibuprofen and propranolol also reacted with chlorine in the experiments performed with San Jose/Santa Clara secondary effluent. However, the transformation was not complete during the one-hour contact time at either dose. Ketoprofen was not removed when free chlorine was added to the San Jose/Santa Clara wastewater effluent. The laboratory data suggest that diclofenac, naproxen and metoprolol react with monochloramine. However, the rates of reaction are relatively slow and probably will not be important under most conditions encountered at municipal wastewater treatment plants. The slower reactions could result in losses of the compounds during long contact times with monochloramine, as would be encountered in a water distribution system that uses monochloramine as a residual disinfectant.

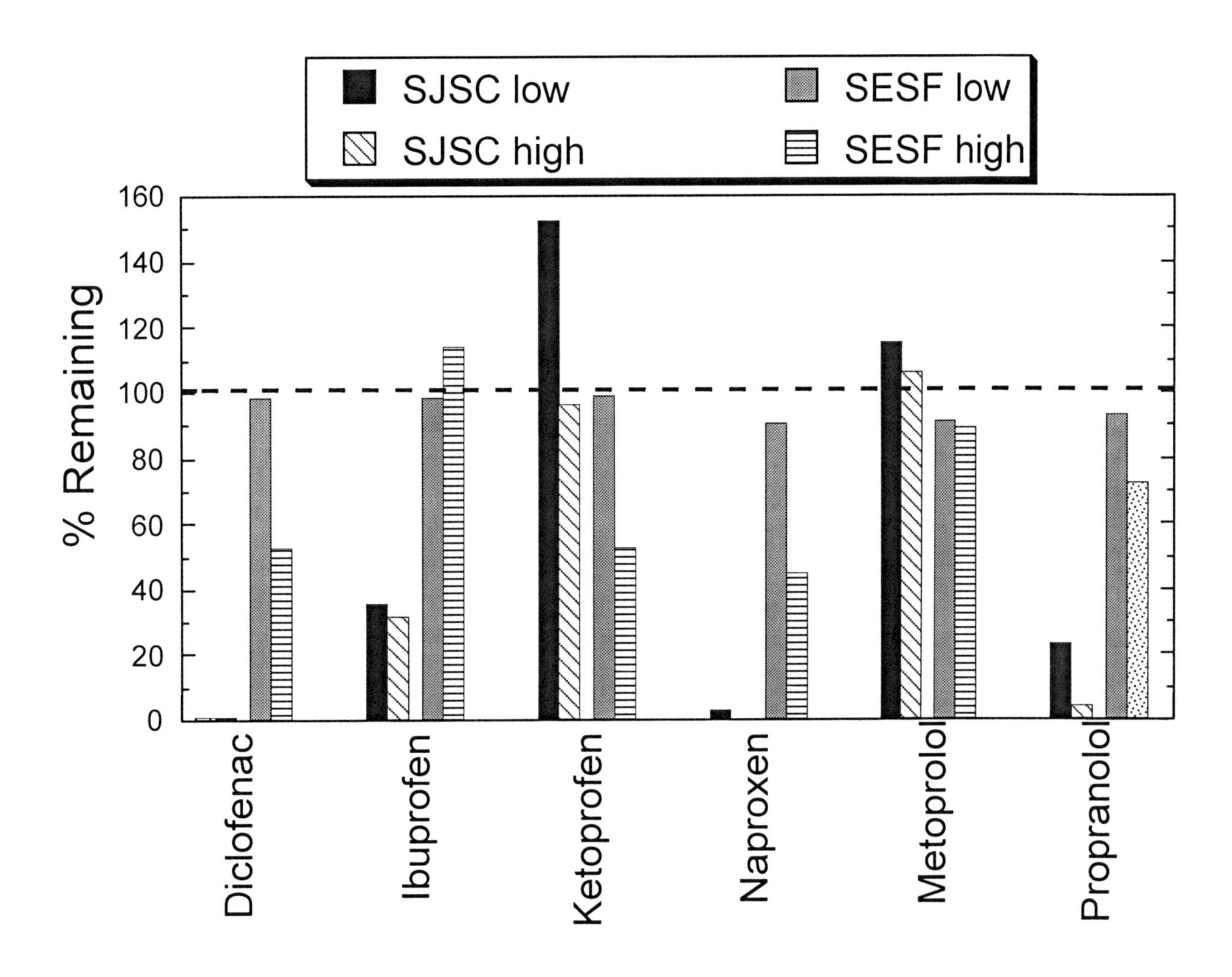

Figure 3.12 Results from laboratory experiments involving the addition of low (10 mg/L as Cl_2) and high (60 mg/L as Cl_2) doses of chlorine to secondary effluent samples from the San Jose/Santa Clara (SJSC) Municipal Wastewater Treatment Plant and Treatment Plant S.

To further investigate the modest decrease in concentrations of pharmaceuticals observed upon exposure to monochloramine another simulated chlorination experiment was performed using effluent form the Treatment Plant S (Figure 3.13). After spiking the samples with 2000 ng/L of a mixture of the PhACs, 10 mg/L as Cl_2 of NaOCl (grey bars in Figure 3.13) was added and the samples were allowed to react for one hour. A separate aliquot was spiked with PhACs and subjected to 350 mg/L as Cl_2 as NaOCl. In the former case, the chlorine was all in a combined form due to the presence of excess ammonium whereas in the latter case, the chlorine was in the free form due to the excess of chlorine relative to ammonium. Analysis of the samples indicated that the concentrations of gemfibrozil, ibuprofen, indometacine, metoprolol and propranolol decreased by 20-50% upon exposure to the low chlorine dose. These results suggest that disinfection with chloramines probably will not have a significant effect on concentrations of these pharmaceuticals. In contrast, the addition of 350 mg/L as Cl_2 of NaOCl resulted in the complete removal of all of the pharmaceuticals except ibuprofen and ketoprofen. Although these are extreme conditions compared to those encountered in disinfection systems, these results further confirm the potential importance of free chlorine disinfection in the removal of pharmaceuticals.

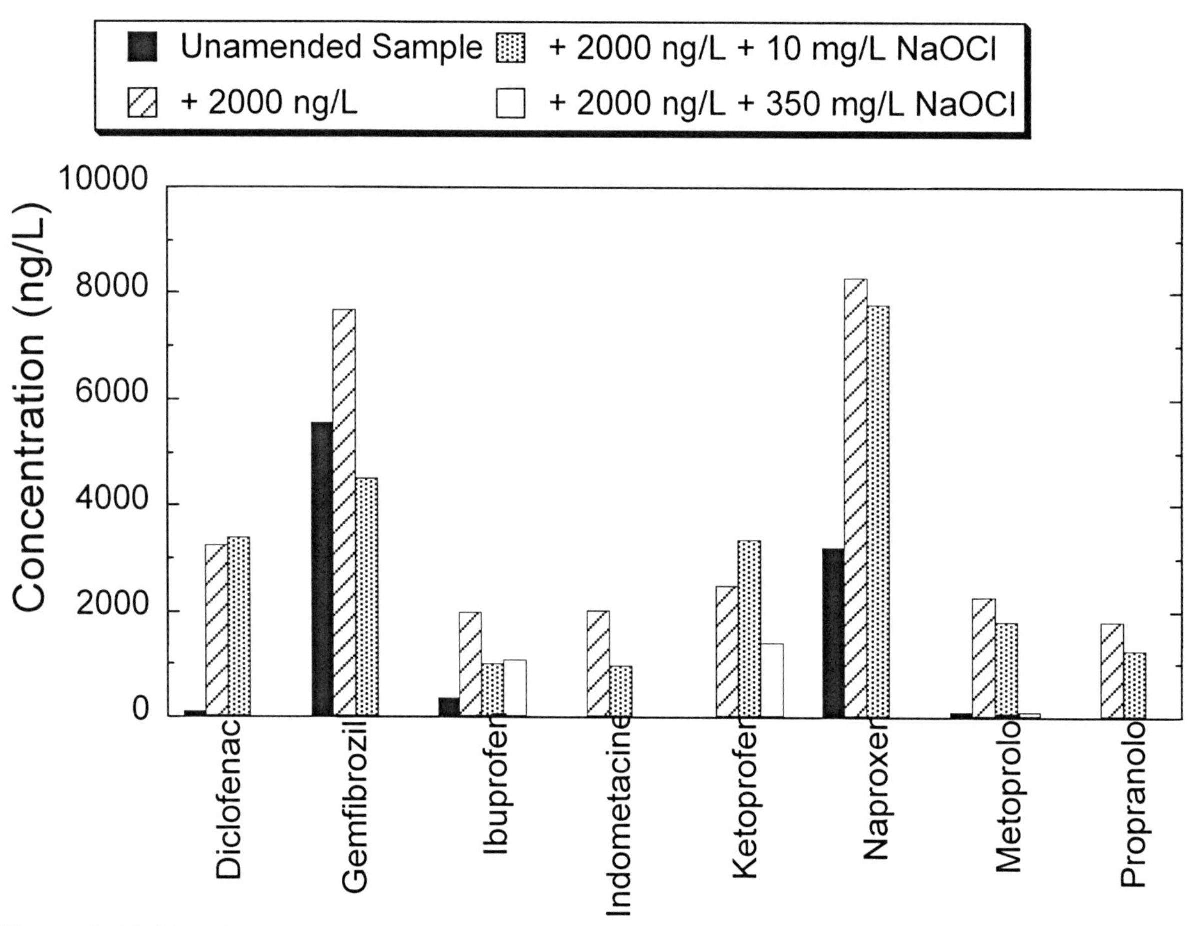

Figure 3.13 Results from laboratory experiments involving the addition of low (10 mg/L as Cl_2) and high (350 mg/L as Cl_2) doses of chlorine for one hour to secondary effluent samples from the treatment plant S. Samples were amended with 2000 ng/L of each pharmaceutical prior to chlorination.

CHAPTER 4
SUMMARY AND CONCLUSIONS

SUMMARY OF FINDINGS

The results of this research project contribute to a better understanding of the occurrence of PhACs in potential drinking water sources. Results of the literature review helped to define the universe of PhACs that could be detected in the United States using readily available analytical methods. The method development and testing activities provided a robust and sensitive means of detecting a suite of target compounds by GC/MS/MS and HPLC/MS. These methods then were used in a limited occurrence survey that quantified PhACs in potential drinking water sources and evaluated the potential for removal of the compounds in different types of treatment systems. Major findings are summarized below.

A review of the available scientific literature indicated that the main source of PhACs in surface and groundwater is municipal wastewater. In most cases, the wastewater has undergone treatment prior to discharge, in which case, some of the PhACs may be removed in the wastewater treatment plant. Analysis of prescription data in the US is complicated by a lack of information on the exact quantities of active ingredients in each prescription. However, the analysis suggests that hundreds of PhACs could be present in untreated wastewater at concentrations that are detectable with modern analytical methods (i.e., above 10-50 ng/L). Comparison of prescription data with similar types of information from Germany, where much of the available occurrence data for PhACs originates, indicates that the compounds present in the US differ somewhat from that in Germany. As a result, while it may be appropriate to monitor many of the PhACs that have been reported to occur in Germany, monitoring of other PhACs also is appropriate.

Evaluation of candidate analytical methods indicated that the most appropriate techniques are gas chromatography/tandem mass spectrometry (GC/MS/MS) and high performance liquid chromatography/mass spectrometry (HPLC/MS). Previously developed analytical methods can be modified to measure many of the most prevalent PhACs expected to occur in the US, but great care needs to be directed towards method implementation and QA/QC. With appropriate care, acceptable results can be obtained by analysts trained in the use of GC/MS/MS and HPLC/MS systems.

Results from the occurrence survey indicated that PhACs are present at detectable concentrations in the effluent of conventional municipal wastewater treatment plants (Table 4.1). After excluding data that did not meet the QA/QC criteria, the median concentrations of the 15 target PhACs in effluent from conventional wastewater treatment plants ranged from <10 to 1400 ng/L. No discernable trend was evident among the WWTPs and concentrations varied by as much as an order of magnitude between samples collected at the same WWTPs on different days. The compounds detected at the highest concentrations were gemfibrozil, naproxen, sulfamethoxazole and trimethoprim. The least frequently detected PhACs were ketoprofen, enrofloxacin, norfloxacin and sulfamethazine. These results are qualitatively consistent with our expectations regarding the use of the different drugs (e.g., enrofloxacin and norfloxacin are not

Table 4.1
Summary statistics for concentrations of PhACs detected in wastewater effluent samples

Compound	Classification	Median	Range	n	% ND*
Ciprofloxacin	Antibiotic	170	<30-860	8	50%
Diclofenac	Acidic	60	<10-78	8	12%
Enrofloxacin	Antibiotic	<60	<30-350	8	75%
Gemfibrozil	Acidic	920	92-5500	8	0%
Ibuprofen	Acidic	50	<10-320	8	50%
Indometacine	Acidic	15	<10-36	8	50%
Ketoprofen	Acidic	<10	<10-55	8	75%
Metoprolol	Beta-blocker	56	9-160	12	0%
Naproxen	Acidic	300	100-3200	8	0%
Norfloxacin	Antibiotic	<60	<30-150	9	78%
Ofloxacin	Antibiotic	180	<30-600	7	14%
Propranolol	Beta-blocker	19	<5-53	12	25%
Sulfamethazine	Antibiotic	<30	<30-500	5	80%
Sulfamethoxazole	Antibiotic	1400	<30-2000	6	17%
Trimethoprim	Antibiotic	550	<30-1900	8	12%

*percentage of effluent samples in which the sample was not detected.

used in human therapy, whereas gemfibrozil, naproxen, sulfamethoxazole and trimethoprim are ranked among the top 30 drugs with respect to predicted concentrations in wastewater).

Analysis of samples from three full-scale advanced wastewater treatment plants indicated that concentrations of the PhACs decrease during advanced treatment. Although microfiltration does not result in removal of the PhACs, application of reverse osmosis after microfiltration lowered the concentrations of acidic drugs and antibiotics below the detection limit. The concentrations of beta-blockers also decreased during reverse osmosis treatment. However, metoprolol and propranolol were detected in one of the three reverse osmosis samples. Concentrations of acidic drugs and beta-blockers were below detection limits after treatment by GAC while several antibiotics were detected after GAC treatment in one of the two samples analyzed for antibiotics. Unfortunately, acidic drugs and beta-blockers were not analyzed in the GAC effluent sample that also contained antibiotics and it is impossible to determine if this was an isolated occurrence. Additional testing may be needed to assess the removal of PhACs during full-scale GAC treatment.

Soil aquifer treatment resulted in a decrease in the concentrations of PhACs. The concentrations of acidic drugs were below detection limits in the shallow well at a soil aquifer treatment site. Concentrations of beta-blockers and antibiotics also decreased during soil aquifer treatment, but metoprolol, propranolol, sulfamethoxazole and trimethoprim were detected in a deep downgradient well. These results are consistent with findings reported by Drewes and Heberer (2002), that indicated the removal of acidic drugs but not the PhACs carbamazepine, phenazone and propyphenazone during soil aquifer treatment at the same location.

Analysis of samples from two engineered treatment wetlands suggested that some removal of acidic drugs and antibiotics occurs. However, the removal is not complete and PhACs can be detected in the effluent of the treatment wetland. Furthermore, the presence of metoprolol and propranolol in water from the Santa Ana River before and after the Prado engineered treatment wetland suggest that the beta-blockers are relatively stable in surface waters.

Transformation of certain PhACs can occur during chlorine disinfection of wastewater effluent, especially when the wastewater has been nitrified prior to disinfection. Results from laboratory experiments indicate that commonly detected PhACs such as diclofenac, ibuprofen, and propranol readily react with free chlorine under the conditions encountered in wastewater disinfection systems. However, attempts to verify the laboratory experiments by sampling before and after wastewater disinfection did not show any clear evidence for removal of these compounds. Removal of the compounds during disinfection with chloramines also is possible, but the rates are too slow to be of significance under the conditions encountered during wastewater treatment.

RECOMMENDATIONS FOR FUTURE RESEARCH

Additional research is needed to develop a better understanding of the occurrence and fate of PhACs in the aquatic environment. Results of this study indicated several areas in which further research is particularly relevant:

- Development of additional simple and robust analytical methods for quantifying PhACs by GC/MS/MS. The use of gas chromatography for PhACs will remain an important tool for analytical laboratories that are not equipped with HPLC/MS systems. Therefore, additional methods should be developed for analysis of PhACs by GC/MS/MS. For example, the beta-blocker method used in this study would benefit from the development of a suitable internal standard.
- Development of additional HPLC/MS methods for analysis of PhACs. Although HPLC/MS/MS offers many benefits over HPLC/MS, tandem mass spectrometers are relatively expensive and HPLC/MS systems usually are easier to operate. Research conducted during this study indicated that the presence of organic matter in wastewater effluent extracts often results in poor recoveries due to matrix suppression. The matrix effects can be controlled by suitable selective solid phase extraction methods. Therefore, additional research on the development of better sample extraction or cleanup methods would be advantageous.
- Detailed studies of the efficacy of soil aquifer treatment systems. Results from this study indicate that certain PhACs are not completely removed during SAT. However, each SAT system is different and it is possible that the removal of PhACs will vary between systems. Additional information is needed on the processes responsible for attenuation of PhACs during SAT as well as approaches that could be used to enhance the efficacy of SAT systems.
- Analysis of the potential for removal of PhACs during chlorine disinfection. Results of this study indicated that certain PhACs may be transformed during when free chlorine is used for disinfection in wastewater treatment plants. Additional research is needed to assess the kinetics and mechanisms of these processes. Additional research also may be merited to assess chlorine disinfection in water treatment plants as well as the fate of the transformation products in the aquatic environment.

- This research project was focused on drugs used for human therapy. Additional research may be needed to assess PhACs used in veterinary applications and personal care products as well as metabolites of human pharmaceuticals.

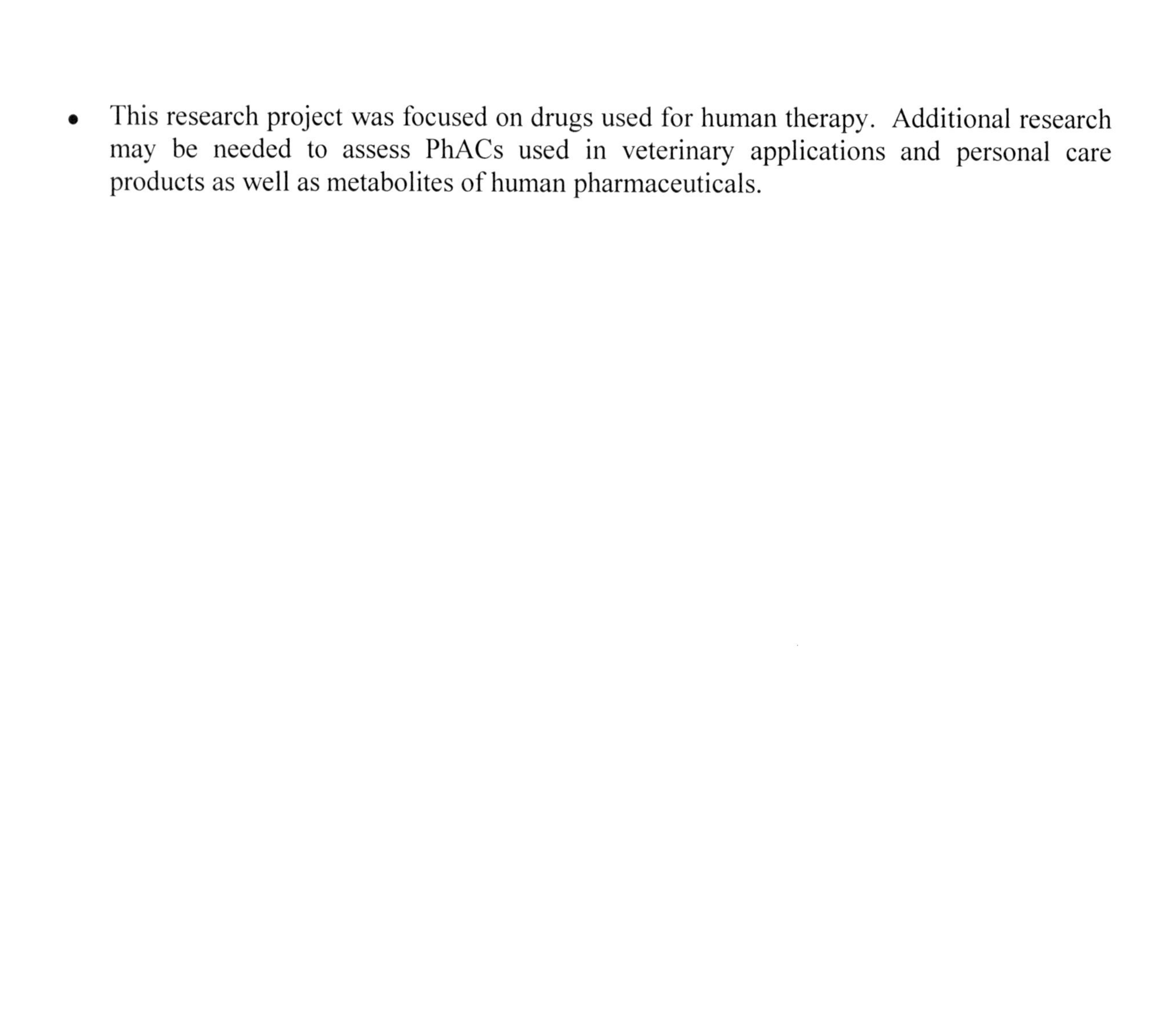

CHAPTER 5
RECOMMENDATIONS TO UTILITIES

As indicated by the research described in this report, PhACs are present at detectable concentrations in municipal wastewater effluent. As a result if intentional and unintentional reuse of wastewater effluent, PhACs often will be present in potential drinking water sources. At present no known human health effects have been associated with exposure to the extremely low concentrations of PhACs detected in the aquatic environment. However, as mentioned in Chapter 1, recent findings of endocrine disruption in fish exposed to trace concentrations of hormones have focused attention on the potential effects of exposure to trace concentrations of pollutants. Although it may take decades for a scientific consensus to arise on the potential threats posed by PhACs in drinking water sources, it is an issue that should be considered in the drinking water community, especially in situations in which indirect potable water reuse is practiced.

Given the uncertainties associated with PhACs and the absence of any specific regulations, it is not appropriate for drinking water utilities to alter treatment practices or to abandon source waters that meet regulatory guidelines. However, drinking water utilities can take a series of steps to anticipate future developments associated with this issue. The extent these actions should depend upon the concentrations of PhACs expected in the utility's source waters. Therefore, the first step that a utility should take involves an assessment of the potential sources of PhACs originating in wastewater effluent, combined sewer overflows, sewage leaks and agricultural runoff. Such assessments are consistent with sanitary surveys that may already have been conducted to protect the source water from pathogens.

If the assessment leads to the conclusion that PhACs are likely to be present in the source water, a limited monitoring study may be appropriate, with analysis of samples from different source waters and after drinking water treatment. If a monitoring study is conducted, the methods described in this report or other methods that provide adequate QA/QC should be used. The selection of an appropriate laboratory and analysis of QA/QC results is crucial because there are no approved standard methods for PhACs. At present, few commercial or utility laboratories are capable of measuring PhACs at the concentrations expected in source waters. Therefore, it may be necessary to develop the in-house capability or to support the development of such capability in an independent laboratory.

REFERENCES

Buser H.R., Poiger T., and M.D, Muller 1998. Occurrence and fate of the pharmaceutical drug diclofenac in surface waters: Rapid photodegradation in a lake. *Environ. Sci. Technol.*, 32:3449-3456.

Chemical and Engineering News, November 2000, Washington DC: American Chemical Society.

Daughton, C.G., and T.A. Ternes. 1999. Pharmaceuticals and personal care products in the environment: Agents of subtle change? *Environmental Health Perspectives*, 107(S6):907-938.

Drewes J.E., Heberer T., and K. Reddersen 2002. Fate of pharmaceuticals during indirect potable reuse. *Water Sci. Technol.,* 46(3):73-80.

Field, T. 1999. Personal communication.

EPA. 1996. Fact Sheet. Washington, D.C., EPA, Office of Wastewater Management.

Heberer T. 2002. Occurrence, fate, and removal of pharmaceutical residues in the aquatic environment: a review of recent research data. *Toxicology Letters*, 131(1-2):5-17.

Hartig, C., Storm, T. and M. Jekel. 1999. Detection and Identification of Sulphonamide Drugs in Municipal Waste Water by Liquid Chromatography Coupled with Electrospray Ionisation Tandem Mass Spectrometry. *Journal of Chromatography A* 854: 163-173.

Hartmann, A., Alder, A.C., Koller, T., and R.M. Widmer. 1998. Identification of Fluoroquinolone Antibiotics as the Main Source of Umuc Genotoxicity in Native Hospital Wastewater. *Environmental Toxicology and Chemistry* 17: 377-382.

Heberer, Th. and H.J. 1996a. Occurrence of polar contaminants in Berlin drinking water. Vom Wasser. 86, 19-31.

Heberer, Th., and H.-J. Stan, H.-J. 1996b. Determination of Dichlorprop, Mecoprop, Clofibric Acid and Naphthyl-acetic Acid in Soil at the Low ppb-Level by GC/MS Using Selected Ion Monitoring, J. AOAC Int. 79, 1428-33.

Hirsch, R., Ternes, T., Haberer, K. and K.L. Kratz. 1999. Occurrence of Antibiotics in the Aquatic Environment. *The Science of the Total Environment* 225: 109-118.

Hirsch R., Ternes T.A., Mueller J., and K. Haberer. 1998. Methods for the determination of neutral drugs as well as betablockers and beta(2)-sympathomimetics in aqueous matrices using GC/MS and LC/MS/MS. *Fresenius Journal of Analytical Chemistry* 362 (3): 329-340 Oct. 1998.

Holm, J.V., Rügge, K., Bjerg, P.L. and T.H. Christensen 1995. Occurrence and Distribution of Pharmaceutical Organic Compounds in the Groundwater Downgradient of a Landfill (Grindsted, Denmark). *Environmental Science and Technology* 29: 1415-1420.

Hou, J.P. and J.W. Poole 1969. Kinetics and Mechanism of Degradation of Ampicillin in Solution. *Journal of Pharmaceutical Sciences* 58: 447-454.

Katzung, B.G. 1998. Basic and Clinical Pharmacology. Stamford, CT: Appleton and Lange.

Kolpin, D.W., Furlong E.T., Meyer M.T., Thurman E.M., Zaugg, S.D., Barber L.B., and H.T. Buxton. 2002. Pharmaceuticals, Hormones, and Other Organic Waste Contaminants in U.S. Streams, 1999-2000: A National Reconnaissance. *Environ. Sci. Technol.*, 36, 1202-1211.

Kummerer K. 2001. Drugs in the environment: emission of drugs, diagnostic aids and disinfectants into wastewater by hospitals in relation to other sources - a review. *Chemosphere*, 45: 957-969.

McArdell C.S., Molnar E., Suter M.J..F., and W. Giger. 2003. Occurrence and fate of macrolide antibiotics in wastewater treatment plants and in the Glatt Valley Watershed, Switzerland. *Environmental Science & Technology*. 37 (24): 5479-5486 Dec. 15 2003.

Meyer M.T., Bumgarner J.E., and J.L. Varns. 2000 Use of radioimmunoassay as a screen for antibiotics in confined animal feeding operations and confirmation by liquid chromatography/mass spectrometry. *Science of the Total Environment*. 248 (2-3): 181-187 APR 5.

Overcash, M.R., Humenik, F.J. and Miner, J.R. 1983. *Livestock Waste Management*. Boca Raton, FL: CRC Press.

Physicians' Desk Reference. 1998. 42nd. ed. Oradell, New Jersey : Medical Economics Co.

Pinkston K.E. 2004. Occurrence of pharmaceuticals and their transformation during chlorine disinfection. Ph.D. dissertation. University of California, Berkeley, CA.

Richardson M.L. and J.M. Bowron. 1985. The Fate of Pharmaceutical Chemicals in the Aquatic Environment. *Journal of Pharmacy and Pharmacology*. 37 (1): 1-12.

Schwabe, U. and D. Paffrath. 1999. *Arzneiveordnungs-Report*. Berlin: Springer.

Snyder S.A., Westerhoff P., Yoon Y., and D.L. Sedlak 2003. Pharmaceuticals, personal care products and endocrine disrupters in water: Implications for water treatment. *Environ Eng. Sci.*, 20:449-469.

Stan, H.-J., Heberer, Th., and M. Linkerhägner. 1994. Occurrence of clofibric acid in the aquatic system - Is the use in human medical care the source of the contamination of surface, ground and drinking water? Vom Wasser. 83, 57-68.

Stumpf M., Ternes T.A, Haberer K., Seel P. and W. Baumann. 1996. Determination of Pharmaceuticals in Sewage Plants and River Water. *Vom Wasser*, 86:291-303.

Stumpf, M., Ternes, T. A., Rolf-Dieter, W., Rodrigues, S. V., and W. Baumann. 1999. Polar drug residues in sewage and natural waters in the state of Rio de Janeiro, Brazil. *Sci. Total Environ.* 225, 135-141.

Ternes T.A., Hirsch R., Mueller J. and K. Haberer. 1998. Methods for the determination of neutral drugs as well as betablockers and β_2-sympathomimetics in aqueous matrices using GC/MS and LC/MS/MS. *Fresenius J. Anal. Chem.*, 362: 329-340.

Ternes, T. A. 1998. Occurrence of drugs in German sewage treatment plants and rivers. Water Res. 32, 3245-3260.

USDA. 1992. The Agricultural Waste Management Field Handbook. USDA NRCS, Washington, D.C.

USDA. 2000. Feed Yearbook. Washington, D.C.

USEPA. 1996. Fact Sheet. Washington, D.C., EPA, Office of Wastewater Management.

ABBREVIATIONS

AFO	animal feeding operation
AS	activated sludge
AU	animal unit
Cl_2	chlorine
GAC	granular activated carbon
GC/MS	gas chromatography/mass spectrometry
GC/MS/MS	gas chromatography/tandem mass spectrometry
gpm	gallons per minute
HLB	hydrophobic lipid balance
HMO	health maintenance organization
HPLC	high performance liquid chromatography
HPLC/MS	high performance liquid chromatography/mass spectrometry
H_3PO_4	phosphoric acid
HRT	hydraulic retention time
kg	kilogram
L	liter
lbs	pounds
MBTFA	N-methyl-bis(trifluoroacetamide)
MF	microfiltration
mg	milligram
MGD	million gallons per day
MSD	mass selective detector
MSTFA	N-methyl-N-(trimethylsilyl)trifluoroacetamide
MWD	Metropolitan Water District of Southern California
NaCl	sodium chloride
$Na_2S_2O_3$	sodium thiosulfate
ND	not detected
ng/L	nanogram per liter
PAC	project advisory committee
PhAC	pharmaceutically active compound
QA/QC	quality assurance/quality control
RO	reverse osmosis

SIM	selected ion monitoring
SPE	solid phase extraction
UC	University of California
US	United States
USGS	United States Geological Survey
UV	ultraviolet
WTP	water treatment plant
WWTP	(conventional) municipal wastewater treatment plant
μg/L	microgram per liter
μm	micron

Printed in the United States
78518LV00002B/239-252

9 781583 213704